D1425603

100391807

Managing Construction Purchasing

- *Contract Buyout*
- *QA/QC Methods*
- *Negotiation Strategies*

John G. McConville, CCC, CPE

Managing Construction Purchasing

- *Contract Buyout*
- *QA/QC Methods*
- *Negotiation Strategies*

John G. McConville, CCC, CPE

R.S. MEANS COMPANY, INC.
A Southam Company

CONSTRUCTION PUBLISHERS & CONSULTANTS
100 Construction Plaza
P.O. Box 800
Kingston, MA 02364-0800
(617) 585-7880

In keeping with the general policy of R.S. Means Company, Inc., its authors, editors, and engineers apply diligence and judgment in locating and using reliable sources for the information published. However, no guarantee or warranty can be given, and all responsibility and liability for loss or damage are hereby disclaimed by the authors, editors, engineers and publisher of this publication with respect to the accuracy, correctness, value and sufficiency of the data, methods, and other information contained herein as applied for any particular purpose or use.

The editors for this book were Suzanne Morris and Vincent P. Murphy; the production coordinator was Marion E. Schofield. Composition was supervised by Joan C. Marshman. The book and jacket were designed by Norman R. Forgit. Some illustrations by Yelena Daskal.

Printed in the United States of America

10 9 8 7 6 5 4 3 2 1

Library of Congress Cataloging in Publication Data

ISBN 0-87629-316-X

Table of Contents

Acknowledgments

Managing Construction Purchasing is the result of five years of research and effort. I wish to express my sincere thanks to the many contractors, manufacturers, and other individuals who have given freely of their time and knowledge so that this book might be produced. I would like to express my appreciation to my wife, Sharon, for her help in typing and proofreading. I would also like to thank my sons, Ian, Andrew, Reed, and Kyle, for their encouragement. Without the contributions of all these individuals, this book would not have been possible.

John G. McConville, CCC, CPE

Introduction

Managing Construction Purchasing provides the construction professional with a step-by-step approach to the entire construction purchasing and contracting function for both U.S. and international construction projects. This book can be a reference tool for individuals who have any involvement with construction purchasing; it provides detailed and readily usable information pertaining to questions that are likely to arise in the construction purchasing arena.

The purchasing requirements of owner companies, government agencies, construction management organizations, architectural and engineering firms, contractors, subcontractors, vendors, and suppliers as they relate to construction work are functionally similar. Although there may be some unique individual requirements, purchasing activities for all construction projects are similar in many respects. The goal of all construction purchasing professionals is to keep all appropriate parties informed about the current purchasing needs and requirements.

There would be little need for this book if all purchasing and construction professionals had ready access to a comprehensive library of construction-related purchasing reference books and knew where and how to research each specific subject. The main purpose of this book is to provide a comprehensive data source for all issues that pertain to construction purchasing and contracting.

This book is designed to benefit many types of organizations. Government agencies and private businesses should find it a valuable reference, as should general contractors, subcontractors, vendors, and suppliers of construction equipment and materials with annual sales exceeding $0.5 million. The book is also intended for design firms, architects, engineers, and construction managers, as well as for mid-sized and large specialist engineering design procurement and construction management companies working in the industrial/manufacturing market. Because many Fortune 500 and 1000 companies are continually undertaking new construction work on their facilities and plants in the U.S. and abroad, personnel in such companies would also benefit from reading and using this book as a reference source. Finally, foreign construction companies working or considering working in North America or other countries will find this book to be an invaluable tool and resource. *Managing Construction*

Purchasing also addresses the interests of U.S. companies considering or performing construction work or services in the international arena.

This book is diversified, covering all the main construction categories:

- Manufacturing/industrial construction (refineries, steel mills, chemical plants, power stations, manufacturing facilities)
- Commercial construction (light industrial projects, shopping centers, hotels, office developments)
- Civil engineering construction (highways, bridges, harbors, water treatment plants)
- Residential construction (housing, condominiums, apartments, small repairs, remodeling projects)
- Institutional construction (public buildings, government facilities)

For these types of construction, the book covers all the fundamentals of purchasing both bulk materials and engineered equipment. Chapter 1 provides an overview of construction purchasing basics, acknowledging the influences of historical events as well as future challenges and opportunities. Chapter 2 outlines the overall procurement process, emphasizing proper planning and problem solving. Chapter 3 gives a detailed illustration of the structure of the purchasing function, from the personnel involved to recommended storage and data collection systems. Chapter 4 covers the many functions related to purchasing construction materials and equipment – quality assurance and control, inspection, expediting, and transportation – and provides several checklists and sample status reports related to these activities. Chapter 5 addresses the legal aspects of the construction process, giving detailed information regarding the types of construction contracts and subcontracts, factors in the selection of contracts and contractors, negotiating, and the bidding process. Chapter 6 turns the discussion over to international construction purchasing, contracting, and subcontracting, reviewing these issues in light of the benefits and possible complications of working in the international marketplace. Finally, Chapter 7 recognizes the other side of the process: selling construction materials, equipment, and services. It is important for purchasing professionals to understand the background and concerns of their counterparts in the sales arena. Also included in the book are approximately 40 sample purchasing and contract administration forms, a list of further references on purchasing, and listings of professional associations and organizations—both domestic and international—that may be helpful in the purchasing process. Finally, there is glossary of standard domestic and international construction purchasing terms and abbreviations.

The forms and charts found throughout the book can be used as they are or modified to fit any construction purchasing situation.

Chapter 1

Introduction to Construction Purchasing

Chapter 1

Introduction to Construction Purchasing

The cost of materials, labor, and professional services procured through construction contracts and purchase orders represents a substantial capital investment for all parties involved in the construction process. It is vitally important that contracts and purchase orders be negotiated, procured, and executed with great care. Improper methods and substandard purchasing and procurement practices can produce costly delays, loss of profit, and possible litigation. This book is suitable for the novice or experienced construction purchasing professional who would like to improve his or her performance of the various responsibilities involved in construction purchasing and contracting.

Successful construction purchasing and contracting requires in-depth knowledge of a variety of subjects and processes, from the initial planning of a project to the final payment and close-out activity. Some of the major topics this book will discuss include:

- Front-end planning
- Contracting
- Construction purchasing
- Procurement and contract administration
- Purchasing planning
- International contracting and purchasing

Also addressed will be important issues such as:

- Quality assurance
- Quality control
- Expediting
- Inspection
- Subcontracting
- Types of contracts

The book also includes a great number of preprinted forms and letters that will assist the construction professional to manage his or her time.

Construction Purchasing Management: The Basics

The term *construction purchasing management* refers to the purchasing and distribution of all the materials, equipment, and services required for the completion of a construction project. The principal responsibilities of construction purchasing management are listed below:

- Obtain materials, equipment, and services that meet project specifications and construction need dates.
- Secure the best commercial and contractual terms available in the marketplace.
- Identify sources of potential damage or loss of materials, equipment, and services, and institute needed controls.
- Provide efficient, timely, and least-cost transportation, storage, and security for materials, equipment, and services before they are incorporated into the construction project.

To survive and grow in today's business environment, every organization must be competitive. Because the materials and equipment cost of a typical construction project generally ranges from 40% to 60% of the bottom line cost of the project, it is critical that the purchasing management function be performed in a professional and efficient manner. A logical, workable purchasing management plan can address all of the intricate details involved in the buy-out process for the materials, equipment, and services needed for every construction project.

This book provides a broad, basic knowledge of domestic and international purchasing and contracting. It is designed for junior and senior buyers, project managers, contracts managers, procurement managers, and purchasing personnel who are first-time buyers or current purchasers of materials, goods, equipment, and services in the domestic or international construction arena.

Construction Team Members

Construction, particularly the purchasing effort, is essentially a team effort in which each member has an important role to play. Figure 1.1 illustrates the organizational structure of a typical construction project team.

- The **owner** (or client) who commissions the work is responsible for purchasing the services of the design team and construction manager and/or general contractor.
- The **design team** (architect or engineer) is engaged by the owner to design the building or facility.
- The **construction manager** is typically used for mid- to large-sized construction projects to plan the construction effort and coordinate the site operations.
- The **general contractor** performs the required construction work and purchases materials, equipment, and services from vendors, suppliers, erectors, and subcontractors.

All of these team members are involved in the construction purchasing process in some way, either directly or indirectly.

Types of Construction

Construction projects vary in complexity, but purchasing and contracting expertise is needed by all of them. Construction projects are usually classified into five main categories or types, which are listed below.

Manufacturing/Industrial Construction

- Refineries
- Chemical plants
- Steel manufacturing plants
- Consumer product facilities
- Power plants
- Food and beverage facilities
- Automobile manufacturing plants
- Hospitals/Biomedical/R&D

Residential Construction

- Housing
- Apartments
- Residential repairs and remodeling
- Condominiums
- Townhouses

Commercial Construction

- Office buildings
- Shopping malls
- Schools
- Motels/Hotels

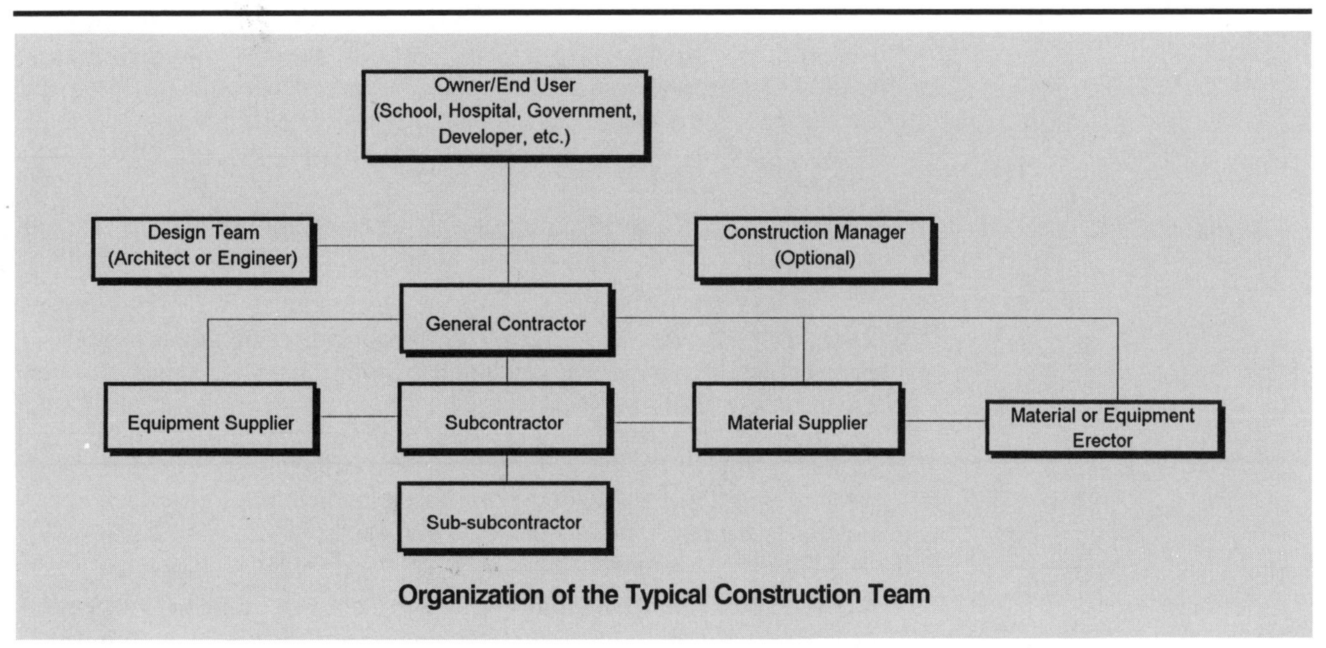

Organization of the Typical Construction Team

Figure 1.1

- Medical offices
- Retail stores
- Warehouses
- Movie theatres

Institutional Construction

- Government buildings
- Libraries
- Hospitals
- Town halls
- Prisons
- Post offices

Civil Engineering Construction

- Bridges
- Tunnels
- Airports (not terminals)
- Water treatment facilities
- Highways
- Harbors/Jetties
- Dams

Costs Related to the Construction Industry

The forecast for construction expenditure in the U.S. for the year 1993 was estimated to be $310 billion. The distribution of these dollars by construction category is depicted in Figure 1.2. (The values depicted are order-of-magnitude numbers, but are reasonably accurate.)

Figure 1.3 illustrates the costs involved in construction on a more detailed level, depicting the breakdown of materials, labor, the design element, and profit on a typical mid-sized construction project.

Cost Savings Potential

Using the information contained in Figures 1.2 and 1.3, it would follow that during 1993 expenditures related to construction materials and equipment in the U.S. would be roughly $155 billion, or 50% of $310 billion. If purchasing management methods, systems, and techniques could be improved to optimal levels, a dollar savings of 5% to 15% could be realized. A savings of 10% would amount to $15 billion in 1993 alone. This sum would accrue as savings and potential profit to owners, contractors, suppliers, and vendors.

The savings of 5% to 15% mentioned above should be within reach of anyone using the techniques and approaches discussed throughout this book.

The Importance of Planning: Two-thirds of all construction projects can be considered small to medium in size. In today's economic environment these projects range in value up to $5 million. Although the methods and systems of construction purchasing management should in theory be the same for any size project, it is reasonable to assume that a more detailed purchasing management approach is necessary for construction projects that range in value from $5 million to $20 million or more.

A construction company may control and execute the construction materials and equipment purchasing effort in many ways – from the

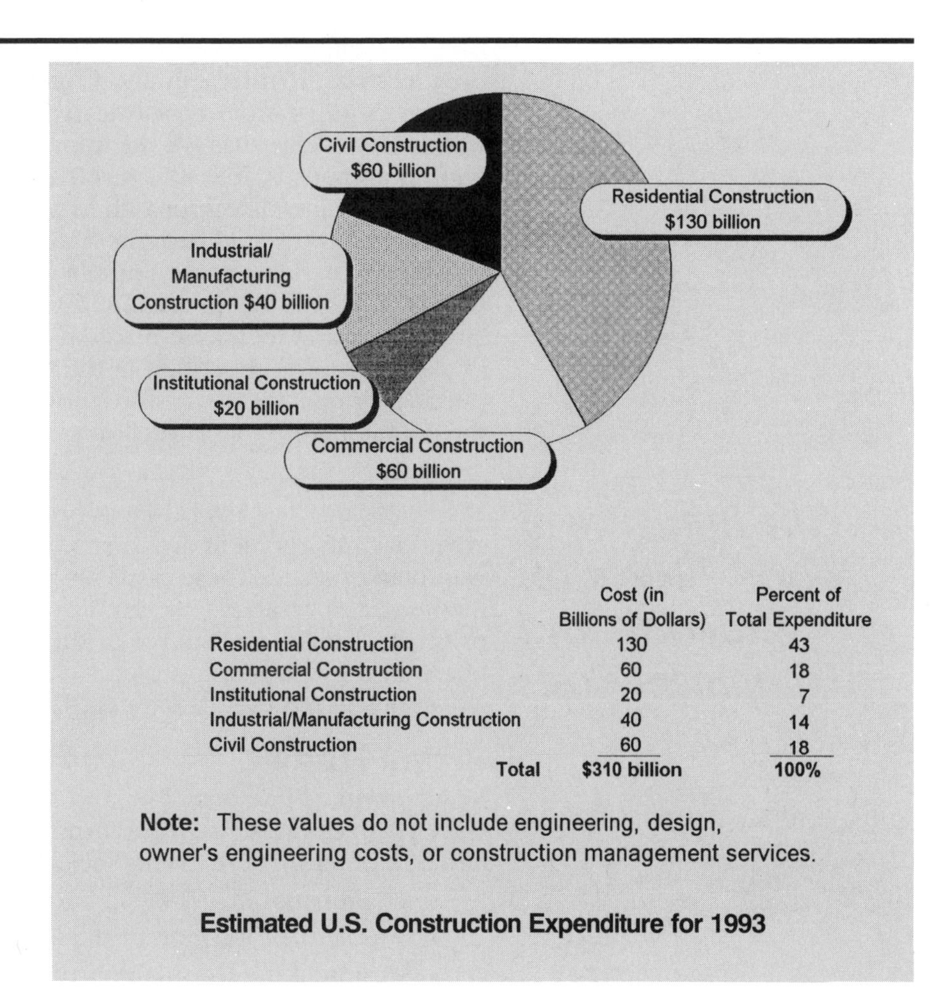

	Cost (in Billions of Dollars)	Percent of Total Expenditure
Residential Construction	130	43
Commercial Construction	60	18
Institutional Construction	20	7
Industrial/Manufacturing Construction	40	14
Civil Construction	60	18
Total	$310 billion	100%

Note: These values do not include engineering, design, owner's engineering costs, or construction management services.

Estimated U.S. Construction Expenditure for 1993

Figure 1.2

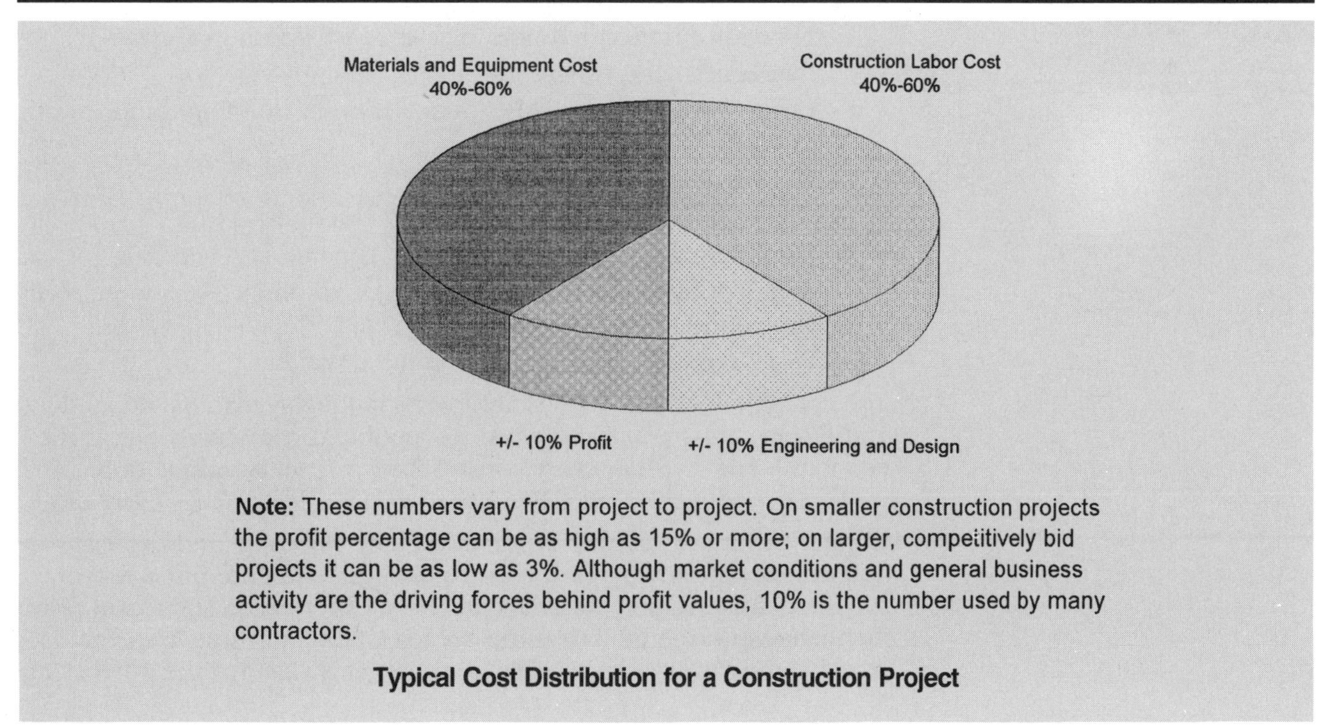

Note: These numbers vary from project to project. On smaller construction projects the profit percentage can be as high as 15% or more; on larger, compeiitively bid projects it can be as low as 3%. Although market conditions and general business activity are the driving forces behind profit values, 10% is the number used by many contractors.

Typical Cost Distribution for a Construction Project

Figure 1.3

back of small contractors' delivery trucks, from a large international Engineering, Procurement and Construction Management (EPCM) organization, or from sophisticated lay-down storage areas or warehouses. Regardless of the approach, the typical cost of materials and equipment is still 40% to 60% of the completed project cost. A planned, businesslike approach to purchasing management can mean significant savings and benefits to the construction project. Having the specified materials and equipment at the correct location at the right time is the recipe for success. Failure to do so results in delays, additional labor costs, and increased materials and equipment costs. Utilization of the techniques described in the following chapters will improve the purchasing management effort and result in increased profits and the containment of construction project costs.

The Influence of World Events on Materials and Equipment Costs

The major goals of any business organization in today's uncertain economic environment are to stay competitive and profitable and to gain market share. These goals are not easily achieved without careful planning and an above average execution strategy. Factors such as price, availability, and source of supplies have an extremely important effect on a construction project – such factors are often influenced by various world events and crises.

Recent History

Many events of the '60s, '70s, and '80s have had a significant and lasting impact on the cost and availability of construction materials and equipment. These events include:

- Oil embargo in 1974
- Double digit inflation of the late '70s and early '80s
- Erosion of the U.S. manufacturing base (many U.S.-based vendors and suppliers are no longer in business)
- Mid-'80s oil price collapse
- Numerous technological changes and advances such as the use of composite materials and plastic construction materials
- Increased market share for Japanese companies
- Large development of ABC (Association of Building Contractors) to promote non-union labor
- Lumber and plywood shortages
- Computerization and automation techniques
- Aluminum and copper shortages in the mid and late '80s
- Ability of countries other than the U.S. to undertake and perform large construction projects worldwide
- Formation of the European Common Market

The construction industry will always remain highly material-dependent and labor-intensive and therefore susceptible to various swings in the economy. The fact that every construction project is unique only underscores the challenges the construction industry faces every day.

Copper products, stainless steel, aluminum, plywood, and the vast majority of petroleum-based construction materials are just a few of the many construction materials that have experienced significant price fluctuations during the last three decades. Some of these materials have been difficult or impossible to obtain. It is vital that the purchasing professional remain aware of situations that could influence the cost and availability of materials and equipment.

Current and Future Events

Major events that will probably influence the availability and cost of construction materials in this decade and the decades to come are:

- A united Europe, which may grow from 12 to as many as 20 member countries
- The opening of Eastern Europe and the significance of their low-cost skilled labor force
- The U.S.-Mexico free trade pact
- The 1991 Gulf War and the large repair effort needed for Iraq's infrastructure
- The rise and fall of Glasnost together with the potentially huge appetite for Western products in the former Soviet Union
- Natural disasters such as Hurricane Andrew; uprisings such as the riots in Los Angeles; the potential for other disasters in this and other parts of the world
- The savings and loan and credit crisis that has put a damper on the residential construction market during the first half of the '90s
- Globalization of the marketplace

Of course, these events have influenced all of society, not only construction materials and equipment. However, the purchasing professional should recognize that some or all of these events can cause delays and/or shortages and increase the cost of materials, equipment, and services. Methods to alleviate some of these potential problems include:

- Close monitoring of the materials purchasing cycle – knowing the location and fabrication status of materials and equipment
- Better forecasting of delays and problems – recognizing potential crises and working around problems to maintain the end date
- Improved communication between all parties involved in the construction process – meetings, status reviews, electronic mail
- Development of workable purchasing management procedures – simple, easy-to-use lists of activities
- More training and seminars in construction purchasing – in-house and external training courses outlining the needs of the purchasing management role
- Computerization of purchasing management systems
- A more professional approach to the overall procurement cycle – hiring bright college graduates, elevating the role of purchasing, joining professional purchasing and related societies

Purchasing and Contracting Communication Problems: Two Case Histories

Construction professionals often encounter challenges in purchasing and contracting. Two specific problems and possible solutions are discussed in the following case studies.

Better Communication, Better Profits

Company A specializes in the fixed price construction of warehouses, office buildings, factories, and various manufacturing facilities. The company is involved in the construction of a warehouse distribution center of approximately 60,000 S.F. for a supermarket chain. The value of the project is approximately $2,500,000.

The estimator employed by Company A is responsible for performing the takeoff of quantities for the underground drains, sewer systems, manholes, and catchbasins, and for obtaining pricing information from local suppliers. He completes a detailed takeoff of the quantities and sends out requests for bids detailing the quantities, specifications, and required delivery time for materials to local suppliers.

After receiving pricing information from suppliers, the estimator incorporates suppliers' material quotations and commercial terms into the bid, using the most competitive prices and commercial terms where possible. He gives this information to the senior estimator, who incorporates all of the information into the overall lump sum estimate. The company submits the bid and two weeks later is successful in securing the job. Company A was the lowest acceptable bid received by the warehouse distribution center owner.

Once the project is underway, the site purchasing agent purchases all the non-subcontract materials. The site quantity surveyor subsequently finds a serious problem. The price the purchasing agent obtained for underground drainage ($150,000) was significantly different from the price contained in the lump sum estimate ($130,000). The $20,000 discrepancy resulted because the purchasing agent had gone out to a completely different group of suppliers. The estimator's efforts to obtain the best material prices were not utilized. A potential profit of $20,000 was lost because of a breakdown in communications.

This situation is a very common one in construction work, often as a result of the involvement of so many individuals and groups in the project.

This type of problem could be prevented by the following measures:

- Distribute supporting documentation for an estimate or proposal to the project team as early as possible in the project's execution phase.
- Have a proposals/estimating coordination meeting with the construction purchasing management group *and* the site staff.
- Distribute a detailed list of estimate or proposal pricing sources to the purchasing management group. This should be accompanied by information on any special discounts or potential bulk purchasing savings, and any special terms or conditions.

In the case of Company A, the problem was not caused by a particular individual – it was caused by improper communication. The project manager should have ensured that every member of his team was aware of the estimate basis. Thousands of dollars could have been saved by improved communication.

Spend Words, Save Money

Problems can also arise when a contractor or subcontractor gives a lump sum price for a particular project without stipulating what is *included* or *excluded* from the total price. A common example is the discovery of rock, unsuitable material, or excessive ground water during an excavation.

The drawings for Building B indicate concrete strip foundations four feet below ground level. When the excavation is done, the city building inspector determines that the excavation material is unsuitable. He instructs the excavation subcontractor to dig an additional two feet. When the cost of this additional work is presented to the client, he refuses to reimburse the general contractor, reasoning that this cost was part of the lump sum bid.

This problem could have been avoided if the bid or proposal letter submitted by the general contractor to the owner had included a brief statement such as, "The lump sum bid is based on foundations being four feet below ground level, with no allowance for the removal of unsuitable material. Any removal of unsuitable excavation material, and any additional concrete and blockwork, will incur additional cost."

It is also advisable for all general contractors, subcontractors, and vendors to stipulate in their bid or proposal the documents and drawings on which the bid is based. It is very important to list the specification number, date, and revision number. The same applies to drawings: the drawing number, date of drawing, and revision designation must be stated and included as part of the bid or proposal. This action establishes the baseline from which the scope of work is determined and on which the price is based. Any changes that are requested by the owner can then be considered as scope changes; either a positive or negative adjustment may be required to the lump sum bid.

A Closer Look at the Importance of Efficient Construction Purchasing Management Systems

The inadequacies of purchasing management methods and systems did not have a serious effect on the overall success of the construction industry until the early '70s. At that time, the Arab oil embargo, the lack of qualified construction purchasing management personnel, and other situations compounded the shortcomings of the purchasing management systems then in place.

Numerous construction experts have carried out a variety of analyses examining how to improve the purchasing management function within the industry. These studies have found that significant money and time can be saved if efficient systems are designed and implemented. These methods should acknowledge the following points:

- Qualified and experienced staff are required.
- For purchase inquiry, bidders must have the necessary documentation along with an adequate scope of work.
- Procedures for evaluating bids technically and commercially must be developed.
- Policies and procedures must be developed for purchasing and negotiation systems, describing functions and responsibilities.
- Quality assurance/quality control requirements must reflect the particular needs of the owner.
- Ensuring that materials and equipment are at the site when needed is vital to the success of any project.

- Transportation and shipping of materials and equipment just in time means that money can be saved by not carrying a large inventory.
- Documentation of the receipt of materials and equipment at the project site must be maintained.
- Materials and equipment must be adequately stored and protected.
- Inventory records and field distribution controls of materials must be established.
- Rework and various related backcharge procedures must be formulated.

The above points relate specifically to the purchasing of materials and equipment but can also be applicable to subcontracting activities. The following points related specifically to subcontracting must also be considered:

- Subcontract planning efforts must be evaluated to ensure the best use of labor.
- Bid packages must be prepared using complete documentation including a list of required deliverables.
- Contractors, subcontractors, and vendors bidding on a job must have the experience and qualifications to perform the work.
- Bidders must be evaluated on a fair and common basis.
- Contractual requirements and obligations must be monitored and controlled.
- All necessary contract close-out activities must be completed.

The Complexities of Modern Construction Projects

The activity chart in Figure 1.4 depicts the various tasks involved in the design and construction phases of a typical construction project. The chart illustrates the complexity of a typical project, detailing the various phases and indicating the parties who are involved.

Certain construction projects are becoming more and more complex, often as a result of circumstances beyond the control of the facility owner or the contracting organization. Fast-track construction projects as well as pharmaceutical, biotechnical, food and beverage, and automated manufacturing facilities require sophisticated instrumentation and control systems. Government agencies such as the F.D.A. and E.P.A. require detailed documentation validating the construction process. This trend appears to be growing.

There are several reasons for the increasing complexity of certain types of projects. State and local government agencies are requiring more permits to cover such factors as wet lands, emissions, and other environmental issues. Government and private companies are performing drug and alcohol tests on their employees, and safety programs are becoming increasingly rigid. In addition, there are rules regarding asbestos removal, hazardous waste, and contaminated soil. After the food poisoning case on the West Coast in 1993, the government is likely to inspect food manufacturing facilities more rigorously, resulting in more inspection of construction methods and more paperwork.

All of the additional paperwork produced by new regulations and permits must be produced, monitored, and adhered to, all of which

Figure 1.4

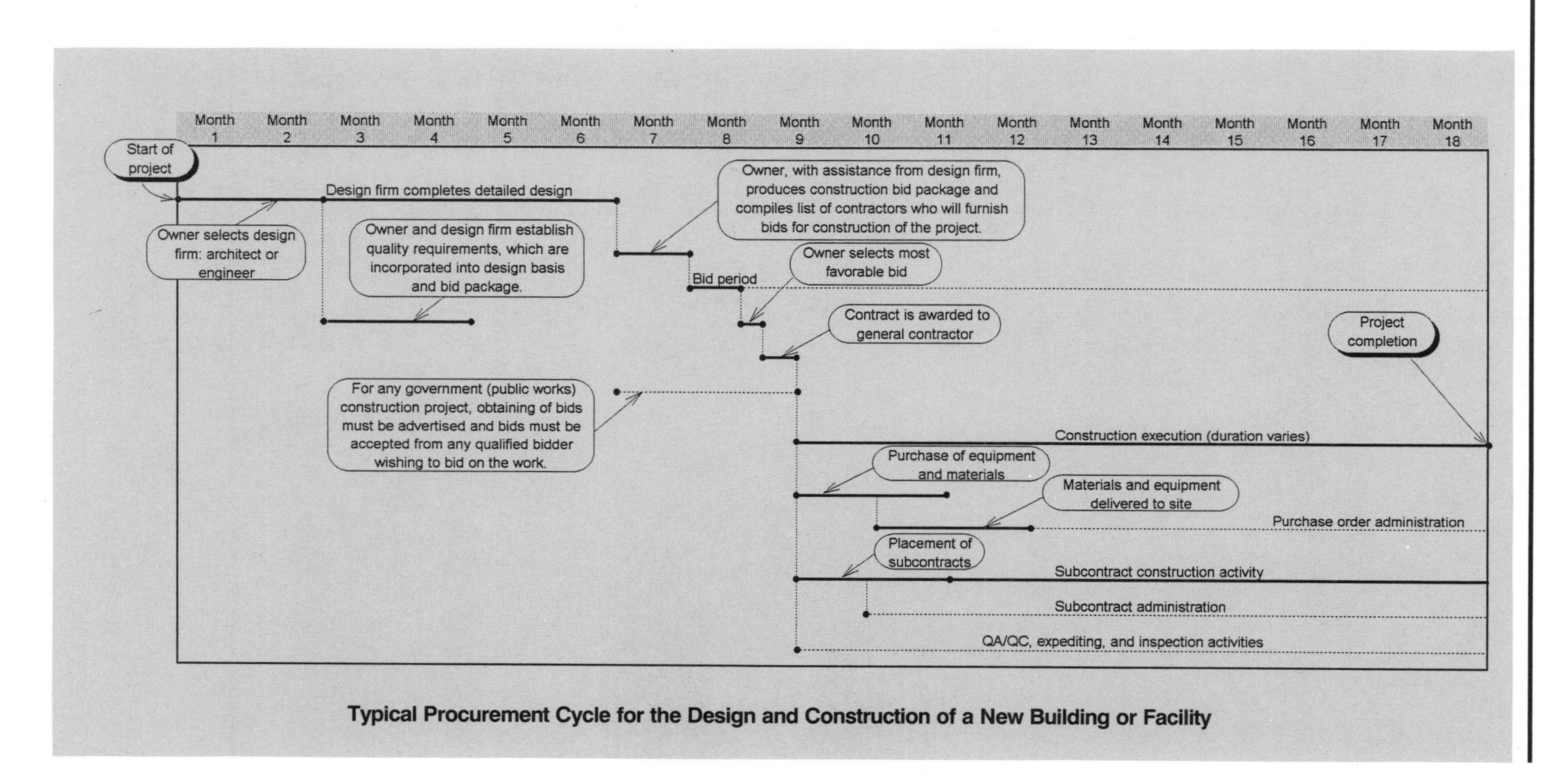

Typical Procurement Cycle for the Design and Construction of a New Building or Facility

require more time and effort. All of these requirements must also be incorporated into bid documents and passed on to contractors, subcontractors, and vendors.

Most of the construction purchasing management systems developed in the '60s and '70s are no longer capable of handling the volume of complex data needed by today's construction teams. The importance of purchasing management has been recognized only recently. Over the last five to ten years, the construction industry has affirmed that improvements in purchasing management have increased productivity both at the home office and on the work site. Significant savings have consequently been realized in the bottom line cost of numerous construction projects.

Challenges to be Faced

Some of the challenges the purchasing professional will face in the '90s and beyond are as follows.

- Training and education will take on added importance in the purchasing management field. Technical advances in equipment, materials, and engineering are being made at a rapid pace, and the purchasing professional will need to stay abreast of future technical advances by attending seminars and courses that relate to purchasing and construction contracting.
- The use of computer technology made rapid gains in the '70s and '80s and it appears that the pace of this technology will significantly increase. Computers are becoming faster, more portable, more user-friendly, increasingly cost-effective, and will influence the purchasing management systems of the future.
- Use of computerized material takeoff procedures for bills of quantities, bills of materials, purchase orders, and inquiry packages will become more commonplace. Computer-aided design (CAD) will become integrated with the materials and equipment takeoff routines.
- As automation becomes more widespread, computers will perform the tasks previously performed by engineers, designers, and purchasing professionals.
- Bar coding will be used more extensively throughout the construction industry, meaning more efficiency in inventory control and ordering and creating significant savings in time and money.
- Quality assurance/quality control programs will become more important in the construction cycle as owners recognize the value of using high quality materials, equipment, and installation methods.
- Greater emphasis will be placed on the communication and coordination effort between design, purchasing, project management, and field construction.
- As a result of limited money and high financing charges, more construction projects will be built using fast-track construction methods, allowing the owner to recover his or her capital investment more rapidly by having the facility in operation sooner.
- The purchasing professional will need to be innovative, looking at new ways to solve purchasing and contracting problems. Projects will need to be examined and reviewed with great care;

- improvements in efficiency and methods will need to be made for the organization to remain viable and competitive.
- Environmental concerns (effects of emissions, asbestos, soil lead, and so on) will influence many decisions and procedures in the construction process, which will in turn have a direct effect on the purchasing management effort.

Purchasing professionals will have a truly unprecedented opportunity to use their talents, knowledge, experience, and professionalism in the exciting '90s and beyond.

International Construction Opportunities

All over the world, authoritarian governments and centralized economies are on the decline while democracies and free market economies are on the rise. This trend became visible in the early '80s and became more and more apparent with the upheavals that occurred in Eastern Europe in the late '80s and early '90s.

As this decade unfolds, East Asian economies will remain a source of expansion and growth. In South America, economic reforms are creating a solid foundation for future development and an improved business environment. The European Community's single market economy will undoubtedly produce economic growth in Europe. The events that have occurred in Eastern Europe and Russia, the problems in the Middle East and the resultant peace dividend, the dramatic increase in the world's population – collectively these events will create major opportunities for profit and growth for construction organizations. There has never been a better time for construction organizations to conduct business overseas.

Globalization

The new buzzword for the 1990s is *globalization*. Globalization can mean different things to different people. As far as this book is concerned it can mean one or all of the following:

- Purchasing a foreign company
- Establishing a joint venture company
- Forming a subsidiary or company in a foreign country
- Performing construction-related activities in an overseas country from a U.S. base

The first step for any construction-related organization considering overseas work is to determine the best option to pursue. There are rewards and risks associated with each choice. Costs, the organization's abilities and experience, and the competition, together with the myriad of foreign government rules and regulations, must be carefully considered and evaluated prior to making the decision to go.

U.S. organizations interested in performing work overseas have a tremendous ally in the federal government. The first action for any organization contemplating overseas work is to contact one of the more than 60 Department of Commerce offices located throughout the U.S. This department can advise on any aspect related to performing work or services overseas. The export-import bank of the U.S. can also be a good source of advice and possible financing. The United States and Foreign Commercial Service (U.S. & F.C.S.) can also provide information regarding overseas work. In addition, the Small Business Administration (SBA) advises small companies on locating overseas markets and on the complexities of working overseas. The addresses of these organizations can be found in Appendix C at the end of this book.

Purchasing a Foreign Company: The purchase of an existing foreign business usually requires a real desire to operate for an extended period of time in an overseas country and, usually, a substantial monetary investment. By maintaining direct or majority control of the business, the owners can compete in a manner that best suits their initial and long-range business and market goals. Many host governments offer some sort of investment or tax incentive that may, in turn, help offset the initial start-up costs of the operation.

Establishing a Joint Venture Company: A wide variety of construction, manufacturing, and service companies maintain joint venture relationships with foreign companies. For some joint venture operations, the host government may require that a majority stake be retained by the local company. Each country has its own unique requirements. That may be an acceptable trade-off for the U.S. firm. The ideal local partner brings a knowledge of the domestic market, valuable business know-how, and very important political and business contacts.

There are, however, potential problems in a joint venture operation. Effective managerial control of the day-to-day running of the business is often lost, and many joint ventures don't last beyond one or two particular projects. The main reason for the breakup of joint ventures is incompatibility of two different management and operational styles.

Forming a Subsidiary or Company in a Foreign Country: Opening a new subsidiary in a foreign country has many advantages. Direct control of business is maintained, and procedures and work practices can be adapted from existing U.S. systems. Local management and staff must be hired and trained, however, and the search for recruitment of these staff members can be expensive and time-consuming.

Performing in an Overseas Country from a U.S. Base: Performing work overseas from a U.S. base is not recommended. Travel costs, travel time, loss of continuity, and a whole host of different work practices, holidays, and cultural differences make this approach unproductive. In addition, clients and foreign employees tend to feel that the organization is not fully committed to long-term business plans in that particular location.

Future Growth Opportunities

The question most frequently asked by executives and managers in the construction industry is, "Where are the best growth and profit opportunities in construction work overseas?" North America, Western Europe, and Japan account for approximately 10% of the world's population but more than 50% of the world's capacity and output. These areas will continue to be large and important markets, offering many opportunities to U.S. construction organizations.

The market areas and countries that are expected to grow rapidly in the next decade or two are:

- Latin America
- Eastern Europe
- Nigeria
- India
- Iraq
- Mexico
- Russia
- Saudi Arabia
- Kuwait
- Indonesia
- China

These countries, as well as many not mentioned, will require substantial amounts of construction work to be performed over the next two to three decades. Their economic development and population increase will fuel this expansion.

Selecting a Country: The second question usually asked is, "Which is the best country for our particular operation?" That depends on the nature of the business, the competition, acceptable risk factors, and a host of other factors. Considerable homework and research must be completed prior to choosing the target country. The topics presented in Chapter 6 (International Purchasing, Contracting, and Subcontracting) will assist in focusing this thought process.

A Closer Look at the New European Community

All across the 12 nations of the European Community the parts of a vast new economic super-power are beginning to fall into place. The long-standing customs, duties, barriers, and tariffs decreased in early 1993 with the formation of a single European market. This event will seriously challenge the international competitiveness and skills of the U.S. construction industry. When trade barriers between the 12 member nations of the European Community (E.C.) are completely dismantled and the 6 members of the European Free Trade Association (E.F.T.A.) adopt parallel codes, practices, and standards, that marketplace will contain more than 350 million consumers, compared with 270 million in the U.S. and Canada.

The Size of the Construction Market: Western Europe had an estimated construction market of $320 billion in 1990, excluding engineering, design, and construction management services and owner engineering services. Figure 1.5 illustrates the size of the construction market in each country of Western Europe.

The Rationale for Interest in Foreign Markets: For some time, the larger U.S. construction-related organizations have developed work abroad in the natural course of expanding their business and markets. Smaller and mid-sized companies also have been performing work abroad – but usually on a very limited basis. The larger construction organizations saw an opportunity to balance the perennial up-and-down swings in the volume of their projected workloads, and for many of them overseas work has become the natural next step in geographical diversification.

Because many construction organizations are having serious trouble finding work in the U.S., interest in foreign markets has never been sharper. This situation coincides with the realization of people in all fields and industries that the United States is not alone, but a part of the global economy.

Research and Planning: The global construction marketplace can be a key to the long-term growth of any construction organization. New customer markets, new technologies, new contractual arrangements, and new purchasing management requirements will require the construction organization contemplating work overseas to do much research and front-end planning. This book sheds some light on this complex undertaking.

Keys to Success

The international construction marketplace is full of tales of disasters encountered by organizations that made basic errors in the execution of their overall overseas implementation plans. Important as it is, construction purchasing management is just one element of the overall implementation plan.

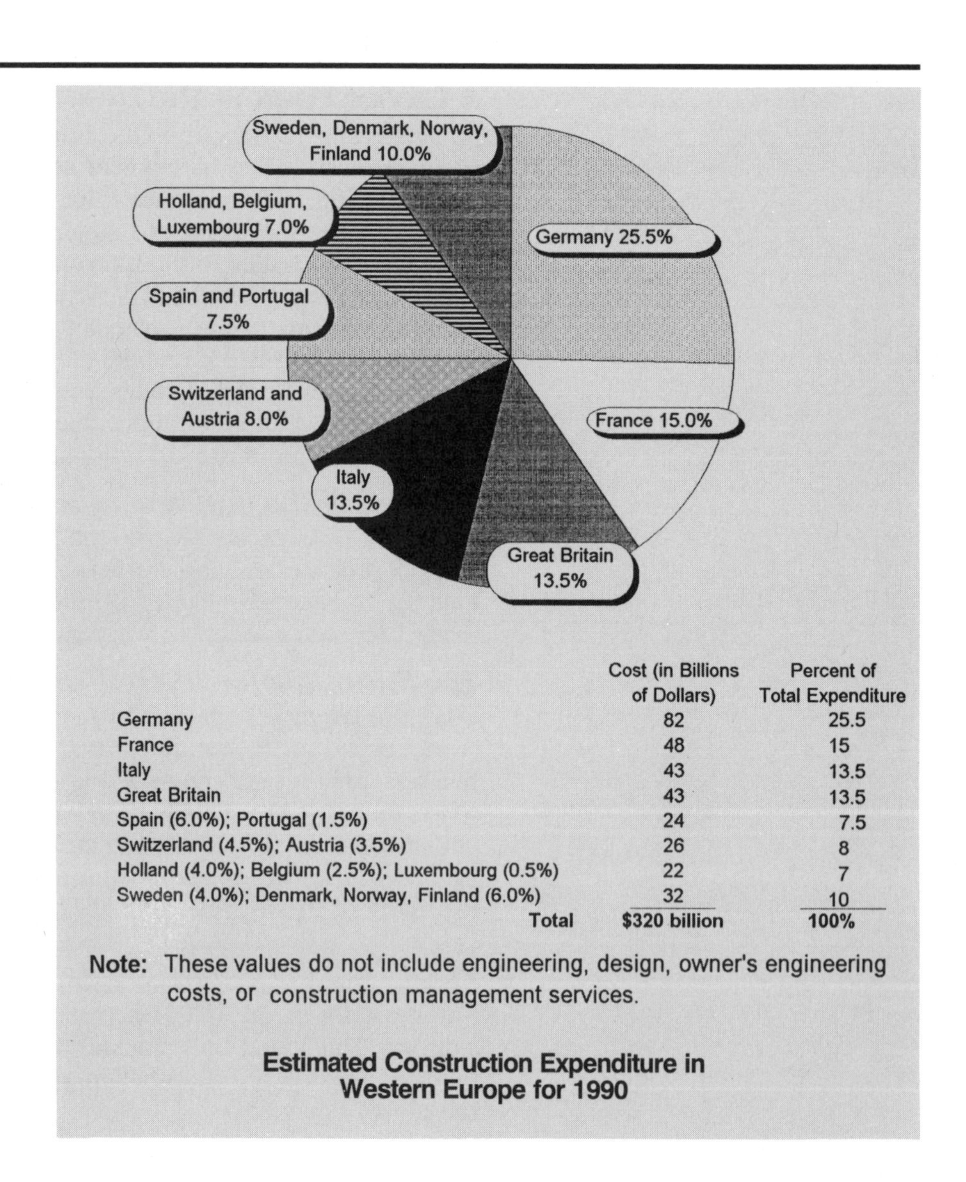

	Cost (in Billions of Dollars)	Percent of Total Expenditure
Germany	82	25.5
France	48	15
Italy	43	13.5
Great Britain	43	13.5
Spain (6.0%); Portugal (1.5%)	24	7.5
Switzerland (4.5%); Austria (3.5%)	26	8
Holland (4.0%); Belgium (2.5%); Luxembourg (0.5%)	22	7
Sweden (4.0%); Denmark, Norway, Finland (6.0%)	32	10
Total	**$320 billion**	**100%**

Note: These values do not include engineering, design, owner's engineering costs, or construction management services.

Estimated Construction Expenditure in Western Europe for 1990

Figure 1.5

Listed below are some of the major do's and don'ts for the construction professional who is involved with international purchasing and contracting.

- Don't underestimate the complexity of the international marketplace. Construction-related organizations have to take a global perspective when setting up an operation overseas. The purchasing management function as it relates to the local market will have to be fully appreciated and understood prior to the commencement of any construction activities.
- Don't assume that by having locally-based foreign nationals working for the foreign operation all of the operational and cultural problems will be solved. Purchasing management personnel from the U.S. domestic operation will have to be sent to oversee and train the foreign purchasing management staff and personnel.
- Do ensure that purchasing management personnel are consulted and involved in the front-end and overall project planning and execution process, including pre- and post-contract negotiations. (This of course applies to domestic purchasing as well.)
- Do ensure that purchasing management personnel are kept informed of any changes to the scope of work of the project. The sooner this group is aware of any changes, the more cost-effective the purchasing and procurement effort will be. (This also applies to work performed in the U.S.)

Chapter 6 includes a more detailed and comprehensive explanation of international purchasing and procurement methods and related activities.

Chapter 2

The Overall Procurement Plan

Chapter 2

The Overall Procurement Plan

Construction projects by their very nature are complex and difficult undertakings. Many different entities are involved, and each activity is dependent on the preceding activity. All activities must be planned and orchestrated so that the project goals and end date are achieved. Construction purchasing management is a major element of any construction project: requirements such as price, quality, and delivery are fundamental elements in the successful execution of any project. Thus *Procurement* is a necessary function in any construction-related organization. The responsibility of the procurement or purchasing management group is to ensure that the correct quantity and quality of materials, equipment, and construction services are at the project location when needed, at a price that is competitive and consistent with the existing economic climate and market conditions. Procurement management maintains control of all suppliers providing all services for the construction project.

Delays in deliveries can be minimized by effective planning and control of the procurement cycle. The procurement planning effort should be substantially more detailed than the typical construction project execution plan. This chapter examines the planning cycles that are necessary to obtain the most competitive pricing levels without sacrificing product quality and risking delays.

The Construction Project Execution Plan

Construction projects typically develop and proceed according to a specific need for a particular building or facility. The owner, often with the assistance of his or her architect or engineer, decides which type of construction contract best suits his or her needs. (Construction contracts are discussed in detail in Chapter 5.) The construction project execution plan must recognize the type of contract the owner wishes to use; the ramifications of this choice may have an impact on the cost and completion date of the construction project.

Many construction projects are delayed because of late delivery of materials, equipment, or services to project sites. Late delivery often occurs because many construction execution project plans use only a single bar line or a single arrow line to designate the procurement function. The description given to these bar or arrow lines is usually "inquire, negotiate, and purchase"(or "I.N.P.") or "purchase and delivery of materials and equipment."

Prior to the start of any construction project, a procurement plan must be integrated into the overall execution plan for the project. Figure 2.1 is an example of an overall construction project execution plan that shows the various procurement, engineering, and construction activities. The document shown is a typical overall project plan related to the construction of a medium- to large-sized manufacturing facility in the United States and shows the various relationships and sequences between engineering, procurement, and construction.

Elements of the Procurement Plan

The procurement or material plan, which is part of the plan depicted in Figure 2.1, must work within the framework of the overall project execution plan. This plan must be based on the overall project objectives, including schedule, budget, quality, location, and any

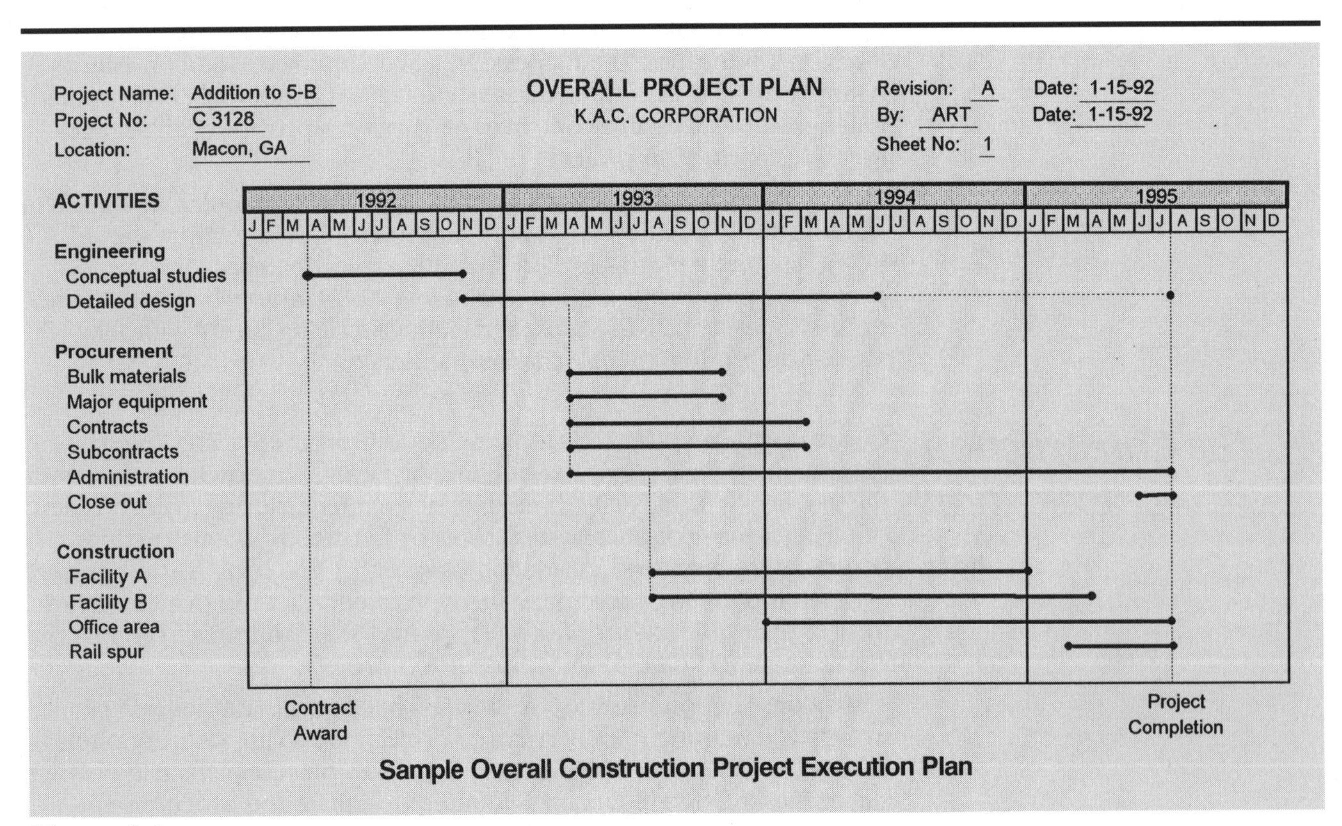

Figure 2.1

particular requirements imposed by the client. Some key questions that must be addressed and integrated into the procurement plan are listed below:

- Which specifications will be used on the project, the owner's or the contractor's?
- Will the owner supply any equipment or materials, or is the contractor responsible for all items?
- Are there any long lead-time equipment items on the critical path of the project schedule?
- What are the quality assurance/quality control requirements of the project?
- Will any equipment or materials be purchased from foreign sources?
- Does equipment need to be furnished with spare parts?
- Is vendor start-up and assistance required?
- What will the specialist subcontractor role be in the buy-out of bulk materials, commodities, and equipment?
- What are the requirements for receiving, storage, and temporary warehousing?
- Can blanket purchase orders for key bulk material items or commodities be placed?
- What are the insurance requirements?
- Are there any special freight requirements?
- Will the home office or site locations be responsible for purchasing?
- How will the project affect existing procurement staffing levels?
- Does the schedule call for a fast-track construction approach or a traditional construction approach?

The above list should be used to generate a more detailed and comprehensive action list. This action list should describe activities, time frames, project restraints, and the personnel responsible for the procurement buy-out and contracting activities.

To ensure that the project is completed on time, it is crucial that responsibility for each activity be assumed by the appropriate party (i.e., either owner or contractor). In the vast majority of projects, the owner or end user of the project will have little or no involvement with the procurement requirements or the overall project implementation plan. The owner is most concerned with the end date of the construction phase, the quality of construction, and the cost of the project.

The Procurement Process

Early planning and communication are the key elements of any successful procurement effort. The discussions and subsequent requirements that result from the above list will assist the procurement effort considerably. The following outline provides a general overview of the main activities involved in the procurement of materials, equipment, and contracts and subcontracts. This is meant to be a step-by-step outline of the main elements of construction purchasing. Details regarding each of the steps will be forthcoming in this book.

Procurement/Contract/Subcontract Planning

- Develop list of suppliers, vendors, and subcontractors
- Develop subcontractor strategy if required
- Define scope of work or services
- Resolve single-source listings
- Resolve delivery requirements
- Review general conditions requirements
- Prepare terms and conditions
- Define insurance requirements

Prepare Bidders Lists/Prequalify Bidders

- List client- or contractor-preferred suppliers or vendors
- List client- or contractor-preferred subcontractors
- List alternate sources
- Develop prequalification questionnaire
- Evaluate responses
- Compile short list of bidders

Prepare Invitation to Bid Packages, Purchase Orders, and Contracts/Subcontracts

- Review specifications and drawings
- Describe scope of work
- Confirm delivery requirements
- Develop bid forms
- Complete instructions to bidders
- Assemble invitation to bid
- Organize pre-bid meetings
- Give brief overview of project (new/renovation, type of building, size, completion date of construction)

Evaluate and Award

- Perform technical and commercial evaluations
- Tabulate bids
- Condition bids for missing scope (additions and omissions)
- Organize pre-award meetings/discussions
- Make recommendation and award contract
- Hold kick-off meeting if required

Contract Administration

- Provide bonds and insurance documentation
- Obtain lien waivers
- Establish payments/backcharges
- Coordinate expediting and shop inspection
- Expedite submittals and deliveries
- Organize meetings and conferences
- Monitor vendors and subcontractors
- Negotiate changes and claims
- Release retention monies
- Complete close-out activities, final reports, and so on

Contracts and Subcontracts

A *contract* is an agreement between two or more parties in which each is bound to perform or forbear some act. A *subcontract* engages a third party to perform work or services, either individually or as part of another's contract.

The Contractor/Subcontractor Relationship

Rarely can one organization undertake the total construction of a building or facility. The most common contractual situation in the construction industry is one in which an owner hires a single organization to handle the project. This organization is usually referred to as a *general contractor* (it may also be referred to as the prime or managing contractor). The owner will look to the general contractor for the overall completion of the project.

Benefits of Subcontracting: When the general contractor can not or does not care to perform certain tasks, he or she will hire *subcontractors* who have the necessary resources, experience, and skills in a particular type of work. These subcontractors usually enter into a legal contract with the general contractor rather than directly with the owner.

A specialist subcontractor can usually perform his or her specialty more efficiently and less expensively than a general contractor could. Economic studies have shown that the subcontract method of construction is cost effective and efficient in terms of both material resources and manpower. By using specialist subcontractors, the general contractor can obtain skilled journeymen with the necessary experience and skills when they are required at the job site. The ideal situation is one in which these specialists are not maintained on the permanent workforce and payroll of the general contractor. More and more general contractors are using specialist subcontractors rather than using their own labor. It is expected that this trend will continue as general contractors endeavor to trim their overhead costs.

Planning for Subcontracting: Subcontractor organizations perform approximately 40% to 75% of all work in the construction industry. The contracting and subcontracting planning process is similar to materials and equipment planning. A contracting/subcontracting strategy must be developed for each area of work; an approved list of contractors and subcontractors must be prepared; terms and conditions must be reviewed and updated; and procedures and standard forms must be revised if necessary.

In many cases, standard terms and clauses of the overall contract between the owner and the general contractor can be passed on to the subcontractor with no direct communication between the subcontractor and the owner. A written subcontract form defines the boundaries of obligation between the general contractor and the subcontractor. This document clarifies and confirms all previous communications, bids, and offers and is proof of both offer and acceptance between the parties.

The Contract/Subcontract Execution Plan

The contract/subcontract execution plan designates the party (owner, general contractor, or subcontractor) responsible for each specific part of the procurement plan. It provides a workable tool for the implementation of procurement activity and defines each party's purchasing responsibility.

The following is a detailed example of a contract/subcontract execution plan that was developed and used on a research and development center expansion project. Included are some of the work packages developed as a result of meetings between the owner, the general contractor, project manager, and engineering, procurement, and construction groups. This list is not intended to be all-inclusive, nor does it apply to every project. Other activities might be required for other projects, and the specific division of responsibility between general contractor and subcontractors may be different for any given project.

Contract/Subcontract Execution Plan

Description of Work	Organization Responsible
Prepurchased Equipment	Purchased by client
Lab equipment	To be installed by GC
Radiation equipment	
Food service furnishings	
Work Package 1. Site Work	To be performed by GC
Scope of Work Included:	
Clearing and grubbing	
Erosion control	
Rough grade	
Excavate for building foundations	
Swales	
Soil removal and disposal	
Site storm drainage and catch basins	
Base paving (partial)	
Dismantle existing warehouse	
Work Package 2. Foundations/Concrete	To be performed by GC
Scope of Work Included:	
Foundations, including forming, rebar, and concrete	
Furnishing and installation of anchor bolts	
Set leveling plates and grouting	
Embedded iron and sleeves	
Slab on grade	
Elevated slabs	
Grounding	
Biowaste pit	
Demolition of building canopy	
Work Package 3. Structural Steel	To be performed by subcontractor
Scope of Work Included:	
Fabrication and erection of structural steel	
Fabrication and erection of decking	
One set of building steel stairs	
Roof screen support steel	
Structural support attachments for dryvit/windows/louvers	

Description of Work	Organization Responsible
Work Package 4. Dryvit	To be performed by subcontractor
Scope of Work Included:	
Fabrication and installation of dryvit panels	
Attachment of dryvit panels to building steel	
Roof screen	
Exterior caulking (panel to panel)	
Flashings at roof and wall	
Work Package 5. Roofing	To be performed by subcontractor
Scope of Work Included:	
Insulation	
Roofing	
Aluminum flashings	
Pavers	
Wood blocking/gravel stops	
Work Package 6. Elevators	To be performed by subcontractor. GC to fit rails, bolts, sills,etc.
Scope of Work Included:	
Furnish and install two elevators	
Drilling of hydraulic piston holes – grouting holes	
Wiring of elevator controls	
Power wiring from disconnect provided in machine room	
Work Package 7. Architectural Interiors	To be performed by GC
Scope of Work Included:	
Glass and glazing (interior and exterior)	
Louvers	
Access doors	
Interior framing	
Drywall	
Ceilings	
Masonry	
Hollow metal doors, bucks, and hardware	
Flooring	
Pass-throughs	
Bumper rails/trims/corner guards	
All interior painting/special guards	
Insulation (thermal and sound insulation)	
Vinyl wall covering	
Bathroom specialties/toilet partitions	
Lockers	
Fire extinguisher cabinets	
Signage	
Millwork	
Interior caulking and weatherstripping	
Fire shutter at "C" building	
Conference room accordion door	
Roll-up doors and motors and controllers	

Description of Work	Organization Responsible
Work Package 8. Underslab Piping	To be performed by GC

Scope of Work Included:

- All underslab piping from termination at 6″ above ground floor to 5′ 0″ outside the foundation wall, including:
  - Storm drains
  - Sanitary drains
  - Biowaste piping
  - Process waste piping
- Excavation and backfill for underslab piping
- Floor drains and clean-outs
- Testing and inspections

Work Package 9. Mechanical Piping	To be performed by GC

Scope of Work Included:

- All yard piping, including:
  - Steam and condensate
  - City water and hydrants
  - Gas
  - Condenser water
  - Cooling tower make-up water
  - Fire water, including hydrants
  - Sanitary and process waste
  - Biowaste pit discharge piping
  - Foundation drain piping
  - Trench drains at loading docks
  - Curb boxes
- Excavation and backfill for yard piping work, including shoring and bracing and disposal of excavated materials
- Thrust blocks
- Demolition of existing steam and condensate and concrete pipe supports serving building "C"
- Sewage ejector at building
- Lift station at road
- Tie-ins of all new services to existing services
- Building piping, including:
  - Steam – high pressure and low pressure
  - Condensate
  - Water, waste, vent, and storm piping
  - Domestic hot water
  - Compressed air – instrument and process
  - Nitrogen
  - Oxygen
  - CO_2
  - Diluted CO_2
  - Gas
  - Reheat hot water
  - Cooling tower water supply and return
  - Chilled water supply and return
  - Glycol supply and return
  - Vacuum
  - Lab piping

Description of Work	Organization Responsible
Roof drains	
Floor drains (second and third floor)	
Hangers, supports, and anchors for building utility piping	
Thermal insulation of building piping	
Provision of valves in utility piping mains for connection by the process piping installer	
Furnishing and setting sleeves	
Cutting patching and sealing	
Testing and start-up	
Steam and water pressure reducing valves	
Equipment Furnished:	Purchased by GC
Sewage ejector pump	
Lift station pump	
Cooling tower water pumps	
Cooling towers	
Reheat coils	
Humidifiers	
Chilled water pumps	
Domestic hot water heater	
Reheat hot water transfer package	
Lab vacuum pumps	
Condensate pumps	
Environmental rooms (Balley Boxes)	
Plumbing fixtures	
Expansion tanks	
Unit heaters	
Perimeter hot water baseboard	
Work Package 10. HVAC – Sheet Metal	To be performed by subcontractor
Scope of Work Included:	
Supply, return, and exhaust ductwork	
Ductwork insulation	
Ductwork hangers and supports	
Vibration isolation	
Duct connections to process equipment furnished	
Hoods, including dust collector and chemical storage hoods	
Duct cleaning	
Ductwork access doors	
Sleeves	
Start-test and preliminary air balance	
Work Package 11. Automatic Temperature Control	To be performed by subcontractor
Scope of Work Included:	
Furnishing and installation of control/monitoring/alarm system	
Furnishing of in-line control installation by others	
Thermostats and humidistats	
Pneumatic tubing	

Description of Work	Organization Responsible
Low-voltage electrical wiring	
Smoke dampers and unit shut-down	
Automatic dampers and actuators	
Furnishing of flow monitors	
Installation of the pressurization monitoring and alarm system	
Room pressure sensors	
Pneumatic tubing installation	
Low voltage electrical wiring	
Software package validation	
Start-up and check-out	
Work Package 12. Process Piping/Instrumentation	To be performed by subcontractor, labor only All materials supplied by GC and issued to subcontractor
Scope of Work Included:	
All process (product) piping	
Connection of utilities to process system equipment from valves left on pipe racks installed by others	
Furnishing and installation of sleeves	
Cutting and patching work	
Hangers and supports	
Thermal insulation of process piping	
Documentation of process piping installation, including drawings and weld inspection reports	
Installation of process utilities, including:	
WFI system	
Clean steam system	
CIP system	
Hot glycol system	
Process vent system	
Passivation and flushing of stainless steel process piping systems, including off-site disposal of rinsate	
Testing and start-up of process piping work	
All process instrumentation work, including:	
Panels	
Wiring	
Tubing	
Load cells	
Actuators	
Control valves	
Transmitters	
Temperature and pressure indicators	
Work Package 13. Electrical	To be performed by subcontractor
Scope of Work Included:	
In-building power distribution	
In-building lighting	
All power wiring to process system equipment	
All site power distribution	

Description of Work	Organization Responsible
Setting sleeves, cutting, and patching for electrical work	
Exit and emergency lights	
Site telephone conduit duct bank	
Building conduit distribution	
All fire alarm system work	
All security system work	
All monitoring system work	
Paging systems	
Clock system	
Disconnects for elevators (in machine room)	
Demolition of overhead power service to research building	
Sprinkler alarm connection – from cooling tower to building	
Grounding cooling tower building	
Work Package 14. Fire Protection	To be performed by subcontractor
Scope of Work Included:	
Building sprinklers – hydraulic design and submission to GC	
Hose cabinets and nozzles	
Siamese connectors, water gongs, etc.	
Sprinkler heads	
Cooling tower sprinkler (and sprinklers in cooling tower building)	
Furnish tamper switches (wiring by elect.)	
Furnish flow switches	
Sprinkler signs	
Work Package 15. Landscaping	To be performed by subcontractor
Scope of Work Included:	
Add top soil	
Seeding and sod	
Shrubs and trees	
Fertilize	
Work Package 16. Temporary Power and Light	To be performed by Subcontractor
Scope of Work Included:	
Temporary building lights	
Temporary building power for construction	
Temporary power for welding machines	
Work Package 17. Miscellaneous Metals	To be performed by GC
Scope of Work Included:	
Process platforms	
Two stairs	
Welded C.I. door frames (including elevators)	
Handrails	
Lintels	
Elevator sill angles	
Bilco pit covers	
Dock bumpers	

Description of Work	Organization Responsible
Work Package 18. Miscellaneous Foundations and Concrete	To be performed by GC

Scope of Work Included:
- Steam trench, including excavation and backfill
- Loading dock
- Substation pad/vault
- Cooling tower sump/foundation
- Sewage ejection pit
- Cooling tower pump hose foundation and slab
- Nitrogen tank pad

The contract/subcontract execution plan can also indicate which procurement group is responsible (i.e., home office or site-based group) for the buy out of subcontracts, materials, equipment, and services. This plan can be modified as situations change; some elements of work may be transferred to other subcontractors as the project progresses. Change orders or scope additions may be added to the plan as well.

The detailed plan above outlines the procurement and labor requirements for each work package. It can be very useful in weekly or biweekly site status meetings; the plan can be updated to reflect work that has been completed. Each party knows what they are responsible for and who they must communicate with. The contract/subcontract execution plan can also be developed using the matrix format described earlier in this chapter.

Details Beyond the Contract/Subcontract Execution Plan: There is no substitute for a detailed document that designates which organization is responsible for each element of work. This is why the contract/subcontract execution plan is so important. Additional procedures that should be followed on each project are listed below:

- All bids or proposals from subcontractors should be submitted in a standard format. Bids or proposals that are incomplete should be revised or not considered.
- All bids and proposals should be submitted to a common place – the general contractor's home office or the site location – at a previously established date and time. If time extensions are granted, all subcontractors should be allowed the same time extension. Bids received after the designated time and date should not be considered.
- In his or her instructions to bidders, the general contractor should include telephone and fax numbers, and the appropriate contact person. A bid quotation form, such as that found in *Means Forms for Contractors*, should be used to record each bid in an organized manner.

- The lowest responsible bid should be accepted. Because contract awards, construction contracts, and subcontracts represent sizable dollar values to any construction organization, it is important that these contracts be procured and negotiated in a professional and businesslike manner. Mistakes, errors, or missing items from a subcontractor's bid will always cost the general contractor additional expense after a subcontract has been agreed to and executed. It is therefore vital that front-end planning be commenced as soon as possible.

Further discussion of contracts and subcontracts is included in Chapter 5 (Contracting, Subcontracting, and Administration).

The Materials Purchasing Responsibility Matrix

Figure 2.2 is an example of an equipment and commodities/bulk materials purchasing responsibility matrix related to the construction of a new manufacturing facility. The matrix was developed by the managing contractor in an early phase of the project. In this example, the various scope packages were still not defined but enough details were known to allow the managing contractor to plan for the construction execution phase.

The contract/subcontract execution plan described earlier in this chapter is an additional step in defining and assigning the procurement responsibility for a project.

Materials Planning

There are three basic categories of construction materials. These categories are listed below along with some of the most widely used materials and equipment that fall into each category.

Commodities or Bulk Materials In many cases bulk materials are manufactured according to specific generic industry standards, specifications, and codes. These items are usually purchased in large quantities.

- Stone, gravel, and sand
- Concrete, ready mixed and site manufactured
- Timber/lumber, plywood, particleboard
- Reinforcement, including rod reinforcement and wire mesh
- Structural steel sections and metal decking
- Brickwork and brick-related products
- Blockwork and block-related products
- Insulation materials
- Roofing materials
- Pipe: concrete, cast iron, carbon steel, stainless steel, copper
- Pipe fittings
- Pipe valves
- Paint and related materials
- Electric cable
- Electric conduit
- Lighting fixtures
- Cable trays

Prefabricated Materials These materials (and sometimes equipment) are usually fabricated outside the job site location. On complex projects, specific facilities may be constructed so the prefabrication work can be performed close to the job site.

- Precast concrete items, wall panels, columns, beams
- Pipe spools 2″ and larger
- Prefabricated structural steel elements
- Process modules and assemblies
- Prefabricated formwork
- Prefabricated rooms/offices
- Control rooms
- Module assemblies
- Prefabricated reinforcement
- Skid-mounted equipment packages

Party Responsible / Equipment and Commodities/ Bulk Materials	Blenders	Packaging Equipment	Pressure Vessels	Pumps	Compressors	Refurbished Equipment	Stone/Lumber/Concrete	Primary Steel	Secondary Steel	Pipes and Fittings	M.C. Valves	Substation	Electric Cable/Conduit	Siding/Roof Materials	Windows/Louvers	Office Upgrade	Change Rooms	Office Furniture	Computer System	Instrumentation	Parking Area	Gate House	Fencing	Storage Tanks	Roads	Insulation/Paint Materials	Railroad Materials	Pipe Tracks
Owner		X				X												X										
General Contractor																												
Home Office	X		X	X	X						X	X							X	X								
Field Office							X								X	X	X				X	X	X		X			
Siding/Roofing Contractor														X														
Structural Steel Contractor								X	X																			
Mechanical Contractor										X																		X
Electrical Contractor													X															
Insulation/Paint Contractor																										X		
Off-Site Tanks Contractor																								X				
Railroad Contractor																											X	

Sample Equipment and Commodities/Bulk Materials Purchasing Responsibility Matrix

Figure 2.2

Consumable Materials These items historically do not represent a large percentage of the cost of the overall construction procurement effort, but every material procurement plan should make an allowance for consumable materials.

- Welding materials
- Welding or cutting gases
- Cleaning materials
- Protection or dunnage materials
- Small tools
- Safety equipment: hard hats, safety glasses, face masks
- Fuel, lube oil/greases
- Cutting fluids
- Packing materials
- Medical supplies
- Field office supplies

Time Factors in Materials Planning

The materials planning effort must be made according to the overall project schedule. Lead times must be allocated within the materials plan for the following main items or tasks:

- Material quantity takeoffs
- Inclusion of technical specifications with request for proposals from suppliers and vendors
- Development of bills of quantities
- Finalization of commercial and contractual terms
- Issuance of the request for proposal inquiry documentation
- Clarification meetings
- Evaluation of bids technically, contractually, and commercially
- Condition of bids for missing scope items
- Purchase order award
- Inspection/monitoring and witness testing activities
- Receipt of materials and temporary storage at job site
- Disposal of surplus or scrap materials
- Procurement and contract administration close-out activities

The materials procurement requirements for any construction project can be a sizable undertaking, requiring experienced and knowledgeable personnel. Planning and communication among project team members will always be the keys to success.

Equipment Planning

A great number of equipment items are purchased for construction projects — both small and large. Such equipment is usually assigned a tag number so it can be readily identified throughout the life of the particular facility. The following are just some of the main items that would fall into this category; a complete list would be much more extensive.

- Pressure vessels
- Pumps
- Compressors and blowers
- Heat exchangers
- Conveyors
- Cranes and hoists

- Elevators
- Towers
- Utility equipment: cooling towers, chillers, transformers, storage tanks
- Material handling equipment
- Boilers
- Instrumentation systems
- Scrubbers
- HVAC equipment

These items may also be referred to as major equipment, minor equipment, or auxiliary equipment, and are usually manufactured by vendors that specialize in one or more particular types of equipment.

Time Factors in Equipment Planning

An equipment list should be developed early in the detailed engineering phase of the project. In many cases, certain sophisticated equipment items will be subject to delayed delivery schedules. Equipment such as compressors, towers, and large pressure vessels can sometimes have a fabrication period of up to 12 months; this must be accounted for in the equipment planning cycle.

Listed below are some important activities that must be evaluated and considered during the equipment procurement effort.

- Receive a detailed equipment list from engineering group
- Receive engineering requirements and specifications for inclusion in request for proposal
- Select possible vendors
- Produce any special terms or conditions
- Detail any performance requirements
- Include relevant quality assurance/quality control documentation and procedure requirements
- Finalize commercial and contractual terms
- List any spare part requirements
- Issue inquiry package
- Hold clarification meetings
- Evaluate bids technically, contractually, and commercially
- Condition bids for missing items
- Review any specific warranties and guarantees
- Receive and approve vendor shop drawings
- Perform inspection/monitoring
- Perform expediting activities
- Receive equipment at job site, along with appropriate shipping or transportation documentation
- Ensure equipment data, operating manuals are in place
- Perform close-out activities

Historical data related to manufacturing/industrial construction projects indicates that approximately 50% of the bottom line construction cost is directly related to equipment. It is therefore vital that equipment procurement planning activities be undertaken in the early phases of a project. Early planning will save money and will also help to ensure that the equipment is at the project site as early as possible.

Because the vast majority of manufacturing/industrial construction schedules are driven by the delivery and installation of equipment items, it is good practice to produce a reverse activity or milestone plan that specifies when particular equipment items must be field installed. This reverse plan can be used as a workable tool that can be monitored throughout the project engineering, procurement, and construction cycles.

Quality Planning

The quality of construction work is influenced by the accuracy of the quality control plan. The statements contained in the quality control plan are arrived at by translating the owner's requirements into project specifications and drawings. Quality must be strived for by all project team members: the owner, engineer, architect, construction manager, general contractor, subcontractors, and all suppliers and vendors.

Establishing a Quality Control Plan

Planning can ensure that quality breakdowns and project delays do not occur. A quality control plan or program should be developed early in the execution planning stage of the project. The plan requires input from the engineering or design team and from the general contractor and subcontractors, and should cover all of the project quality requirements.

Considering the Owner's Needs

The next step in the quality planning cycle is to integrate the owner's requirements and needs into the project specifications and drawings. The engineer or architect must use the latest standard codes and materials that conform to the owner's needs and requirements.

The requirements outlined in the quality control plan are usually dictated by the owner or end user of the project. The owner may be concerned with many issues, including:

- Budget
- Reliability of equipment
- Operating costs: labor, electricity, gas, water
- Interchangeability of key equipment items
- Quality of the end product
- Appearance of the building or facility
- Function of the building or facility
- Safety of the building or facility

Communicating Quality Needs to Suppliers

The procurement group must maintain a comprehensive, current, and accurate list of vendors and suppliers, along with details of past and current performance. Bid lists should be formulated only after past vendor performance has been investigated. The quality needs or unusual requirements of the project must be communicated to potential vendors and suppliers.

Early planning related to these quality considerations is the key to ensuring that the quality objectives of the project are realized. Further discussion of quality assurance and quality control is included in Chapter 4.

Inspection Planning

The term *inspection* refers to the physical examination of materials and equipment to ensure that they comply with the contract, subcontract, or purchase order, the specifications, and any statutory requirements and regulations.

Field inspection and inspection of work performed in a vendor's shop are the two main ingredients in quality control. More specific details regarding these types of inspection are included in Chapter 4.

Establishing an Inspection Plan

The planning requirements for inspection should be considered prior to the design effort or as early as possible in the design phase. This inspection planning effort should include the following steps:

- Recognize which equipment and materials are to be inspected. Create a list of these items.
- Perform vendor shop inspections of equipment and materials when a particular item is very complex in nature or critical to the overall safety of the facility, or when a vendor or supplier is fabricating a critical item for the first time.
- Do not limit vendor shop inspections to big ticket equipment items such as vessels, towers, compressors, and so on.
- Decide whether in-house or third-party inspection services are required.
- Arrange for permanent inspection staff at the construction site if the owner or general contractor requires this level of inspection.
- Allow for coordination meetings at the vendor's shop before the vendor commences work on a particular item. Ensure that the vendor fully understands the quality requirements contained in the specification.
- Allow for various inspection issues such as travel to the vendor's shop, access to the vendor's shop, and the types of forms (test results, inspection reports, and certificates) that will be required from the vendor. Prepare for the paper flow of data to the home office or the site location.

Some or all of these points should assist in allaying potentially serious problems related to the fabrication of equipment and materials. Chapter 4 covers the inspection requirements in greater detail.

Expediting Planning

The main goal of *expediting* on a construction project is to ensure that the appropriate materials and equipment are delivered to the construction job site when required. The expediting function involves the continuous review of the performance of vendors, suppliers, contractors, and subcontractors. It includes reporting the status of all purchase orders and contracts from placement of order to delivery.

Expediting services are inexpensive compared to the costs of potential delays caused by late delivery of key equipment and materials. Bulk materials and commodities such as stone, aggregate, cement, and lumber do not usually require intensive expediting efforts, although this may not be the case in a remote, difficult location.

Establishing an Expediting Plan

Once a purchase order or contract has been awarded, it is vital to the success of the construction project that expediting contact be established, maintained, and monitored with each vendor and supplier. A delivery date listed on the purchase order is not a guarantee that the equipment or materials will be at the site when required. An expediting plan helps to ensure that the promised delivery date is maintained.

The overall project plan and the materials and equipment requirements must be outlined in detail to evaluate the need for expediting services. There are three types of expediting methods or approaches that should be planned for, depending on the scale and nature of the expediting requirements on the project. The three levels of expediting, along with further discussion of the expediting function, are described in Chapter 4.

Transportation Planning

In the construction industry, *transportation* means delivery, shipping, and trafficking of materials and equipment from their point of origin, manufacture, or fabrication to a project construction site. The main objective of transportation is to move the materials and equipment in a cost-effective, safe, and timely manner in accordance with project goals. A transportation plan must be established by the procurement group early in the project procurement cycle.

Listed below are some of the factors that must be addressed and considered when formulating a transportation plan.

- Method of transportation (air, truck, rail, or sea)
- Safe packing of equipment and materials
- Domestic or foreign transportation
- Load restrictions and route restraints
- Tax laws
- Full truck or container loads versus half loads
- Freight forwarding requirements
- Tariffs or import levies and duties
- Materials or equipment lost in transit
- Materials or equipment damaged in transit
- Insurance requirements
- Temporary storage facilities
- Off-loading and distribution requirements (cranes, fork-lift trucks, etc.)
- Warehousing requirements

A more detailed description of transportation is included in Chapter 4.

Procurement Administration Planning

The procurement administration planning effort should be formulated well before the issuance of any bidding documents. Procurement management and personnel must have a thorough knowledge of the construction process and of the project goals and objectives as they relate to materials, equipment, contracting, subcontracting, quality, inspection, expediting, transportation, and administration. Poor procurement administration practices can lead to disputes, disruptions, delays, claims, and possible litigation. The subject of procurement administration is described in detail in Chapter 3 (Purchasing).

The procurement administration planning effort should address the topics listed below:

- Organization and staff requirements
- Organizational interface between home office procurement group and site procurement group
- Materials requirements (including engineering interface)
- Equipment requirement (including engineering interface)
- Contracting strategy
- Subcontracting
- Quality assurance, quality control
- Inspection requirements
- Expediting
- Transportation
- Warehousing
- Close out

The Procurement Flow Diagram

The first step in the planning effort is to develop a procurement flow diagram, which is a list of all the various elements involved in the procurement effort. Each project can tailor the diagram according to its unique needs and requirements. The diagram should indicate which party is responsible for each procurement-related function.

Some of the tasks listed below should be integrated into the procurement flow diagram, which will then be incorporated into the procurement administration planning effort.

- Develop list of vendors and subcontractors
- Incorporate engineering specifications and data into purchase orders and subcontracts
- Prepare purchase order and subcontract packages
- Make commercial evaluations
- Review and audit invoice payments and backcharges
- Collect and distribute shop drawings and vendor data
- Collect packing and shipping documents
- Maintain and collect test reports
- Maintain inspection reports
- Maintain progress reports
- Collect operating manuals
- Monitor insurance and bonding information
- Prepare release liens
- Issue retention release
- Review and audit change orders and claims
- Collect and maintain as-built drawings
- Prepare close-out reports and related documentation

The procurement flow diagram (a sample is shown on the next page) should be analyzed to determine whether a more efficient approach could be utilized. Once the final project procurement flow diagram has been established, it should be developed into a working procedure outline and distributed to all appropriate project team members. The achievement of project goals depends heavily on the planning of the entire procurement effort. The procurement plan is as important as the engineering and construction planning efforts; it covers all phases of the project, from the procurement of equipment to the manufacture and delivery of equipment to the job site. Equipment will always remain the most costly, complex, and difficult item to procure. Although materials, equipment, and subcontract planning will be different for each project, certain planning functions remain constant.

This chapter is deliberately conceptual and general in nature, as its purpose is to provide basic guidelines for the procurement effort. A project procurement plan will of course be very specific and should recognize complex factors such as client communication requirements, existing facilities, location of the project, contracting methodology, facility start-up dates, and any other project-specific requirements that will influence the overall procurement plan.

Activity	Initial Responsible Group	Follow-up Activities Responsible Group
Develop list of vendors/ subcontractors	PM, E, P, C	______
Incorporate engineering specifications and data into purchase orders and subcontracts	PM, E, P	______
Preparation of purchase order and subcontract bid packages	PM, P	______
Receive bids related to purchase orders and subcontracts	P	______
Perform technical, contractual commercial evaluations	PM, E, P	______
Make recommendations and award contracts	PM, E, P	______
Review payment requests	P	______
Monitor and approve changes	PM, P	______
Expedite shop drawings and other required submittals	E, P	______
Collect transportation related documentation	P, CS	______
Maintain test/witnessing data and reports	E, P, CS	______
Collect and distribute various operating manuals	E, P, CS	______
Maintain insurance and bonding status file	P	______
Purchase order and contract administration activities	P	______
Collection of as-built documentation records and data	P, CS	______
Final close-out activities, reports, data collection, audits, final payments and release of retention monies, and completion certificates	PM, E, P, CS, C	______

The procurement flow diagram shown above indicates the various project groups that are involved with procurement activities. The individual groups are indicated as follows:

PM *project management group*
E *engineering group*
P *procurement group*
CS *construction site group*
C *client*

Chapter 3
Purchasing

Chapter 3

Purchasing

Purchasing refers to the acquisition of materials, goods, and services, and the establishment of mutually acceptable terms and conditions between the seller and the buyer. Purchasing is a fundamental function of any construction-related organization. The purchasing group is responsible for having the correct quantity and quality of materials, equipment, and services on hand at the project construction site when they are required at a price that is consistent with the established project budget.

Although the specific range of responsibilities may vary for individual companies, the essential duties of the purchasing group are:

- Identify and select possible vendors, suppliers, and subcontractors.
- Prepare and issue requests for bids and subcontracts.
- Receive bids and proposals from vendors, suppliers, and subcontractors.
- Perform commercial evaluations of bids and subcontracts including conditioning of bids and subcontracts for noncompliance.
- Undertake and satisfactorily conclude negotiations with potential vendors, suppliers, and subcontractors.
- Coordinate committed purchase orders and subcontracts, including any applicable change orders or backcharges.

These operations should be performed in a professional and ethical manner consistent with the procedures and policies established by the contractor and, in some situations, the owner. The purchasing group's mission is to obtain goods and services from sources that provide acceptable quality at competitive pricing levels. Other aspects that must be considered are compliance with project specifications and delivery requirements, warranties and performance requirements, and supplier or vendor assistance during start-up of the facility or building.

Organization and Structure

Regardless of the size, scope, or type of business entity, a number of fundamental guidelines are essential in developing a sound and successful organizational structure. These principles apply to any business entity – be that entity a small repair and remodeling contractor, a government agency, or one of the largest engineering procurement and construction management firms in the world. Size or financial clout has no influence on this standing. Three fundamental principles for any successful organization follow.

1. A business entity's overall structure or organization should be determined only after consideration is given to the goals and the specific mission statement of the business.
2. A company's resources should always include experienced, knowledgeable, and professional staff.
3. The delegation of authority in a business should be readily understood and facilitated by pre-established lines of authority and communication.

In addition to these three principles, there are many other important considerations in the planning and developing of an organizational structure for construction operations. Some of these important considerations are listed below.

- Size of construction organization.
- Size, type of services, and location of construction projects.
- Authority and responsibility of key staff.
- Home office and site office relationships.
- Communication and interfaces with other operations, groups, or departments within the company.
- Reporting lines.
- Dotted-line administration responsibilities and communication.
- Line and staff functions and roles.
- Status and financial reporting systems.
- Flexibility to allow constant evaluation, change, and improvement.

Public Construction Versus Private Construction

For any government construction project, the law requires that the date for receiving bids be advertised and that bids be accepted from any qualified bidder interested in the work. This is typically the case for city, state, and county projects, which are classified as *public works*. Public works construction is usually financed through public funds and bond issues. The regulations and requirements for public construction work are usually more stringent than those for private work.

For privately owned construction projects, there is also a specified due date for bids. It may not be advertised in trade publications, and the list of bidders may consist of prequalified contracters. Financing is obtained through various private channels.

The operating principles discussed in this chapter and subsequent chapters can be used for both public and private construction work.

Organization of the Purchasing Process

The purchasing function for construction projects is usually managed according to one of three distinct approaches: (1) centralized home office purchasing, (2) decentralized site- or project-based purchasing, or (3) a combination of these two approaches. Both the general contractor and owner must consider project goals in deciding which approach is to be used.

The Typical Project: The basic organization for a typical construction project is depicted in Figure 3.1. The project team is divided into three distinct groups: engineering, procurement, and construction. These groups are usually supported by project controls and home office corporate and administrative resources.

The Home Office Purchasing Department: The construction purchasing organization of a medium-to-large contractor is depicted in Figure 3.2. The chart shows the various reporting and communication lines between the home office and the site-based group. Smaller contractors could tailor this organization to match their specific objectives; for instance, certain individuals could perform a combination of functions.

The organization depicted in Figure 3.3 specifically shows the staffing requirements and reporting lines of the transportation, purchasing, and expediting elements of the procurement function, together with dotted-line administration and responsibilities. Smaller contractors might modify this organization to suit their specific future applications. Certain individuals could wear two or three hats to ensure that all functions are satisfactorily covered. Detailed descriptions of responsibilities for many of the jobs included in Figure 3.3 appear later in this chapter.

The Stand-Alone Domestic or Overseas Construction Project: Figure 3.4 depicts the organization of the typical structure that would be in place in a medium or large stand-alone domestic or overseas construction project. The term *stand-alone* means that the contractor's home office has little or no direct involvement with the construction project.

Figure 3.1

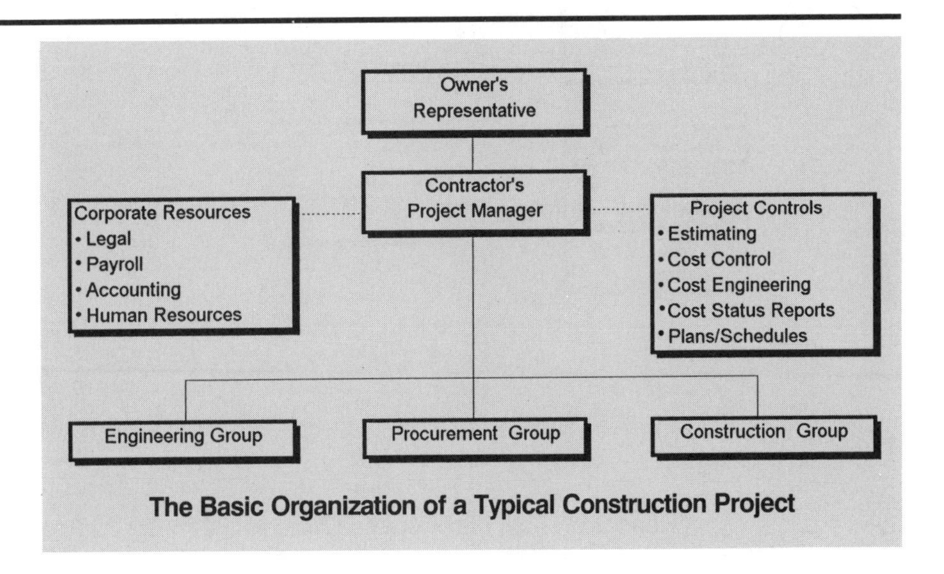

The Basic Organization of a Typical Construction Project

The organizational chart is in a summary form. A more detailed organizational chart would show individual names and reporting relationships for each group. The chart would also depict dotted-line administration responsibilities back to the contractor's home office. Smaller construction projects could also use the organizational concepts indicated. The organizational principles apply to all construction projects, both large and small.

Benefits of Effective Organization

Any business entity can realize the benefits of the three principles of organization mentioned earlier. A company might incorporate such basic principles with structural considerations based on its own individual needs and plans. Project objectives will be achieved in a more effective and efficient manner and the day-to-day operations of the business will be greatly enhanced. Listed below are some of the main advantages of effective organization.

- Use of organizational chart to assist management in the distribution of work tasks or assignments, preventing possible duplication of work.
- Use of organizational chart to allow easier communication between various groups and individuals.

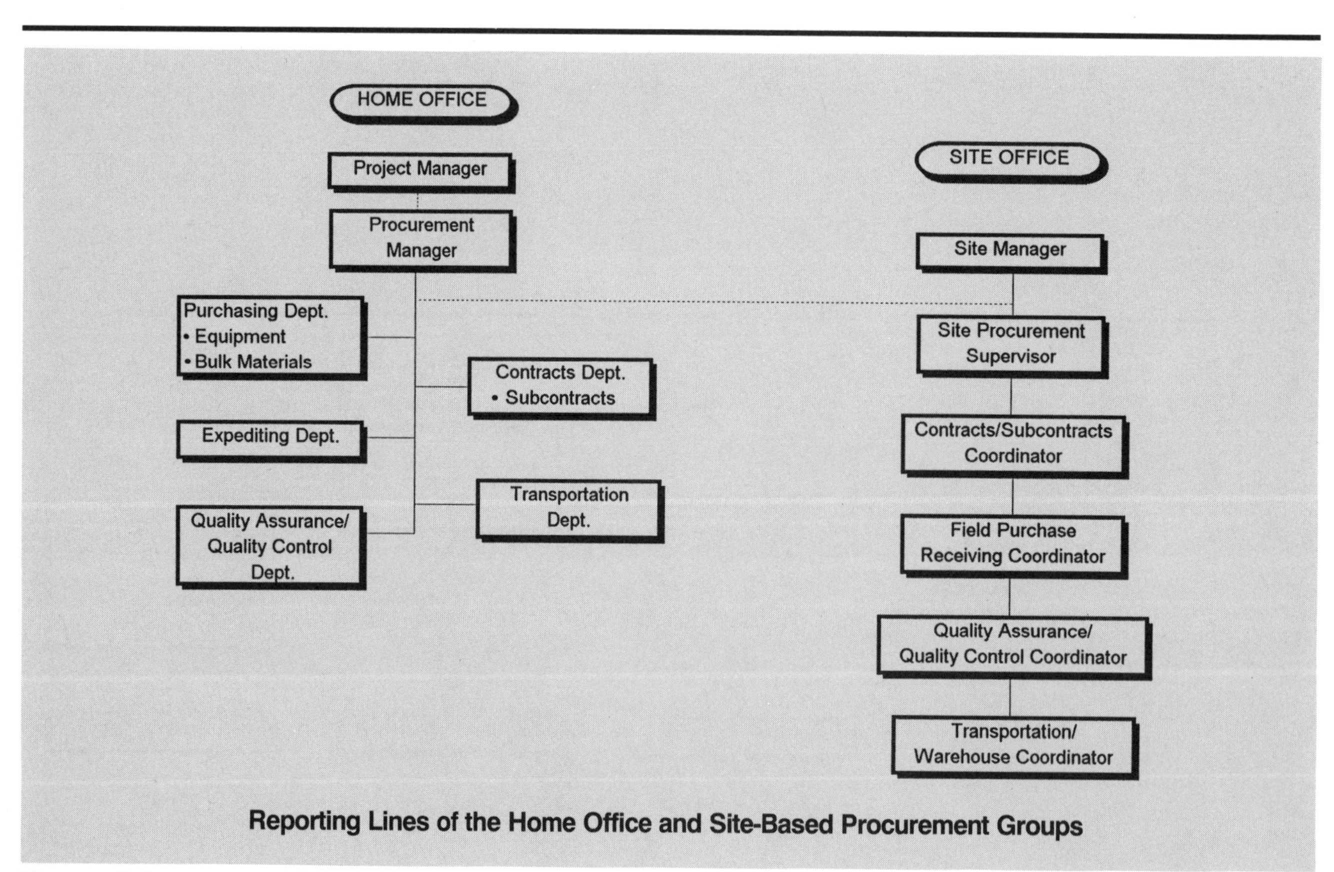

Reporting Lines of the Home Office and Site-Based Procurement Groups

Figure 3.2

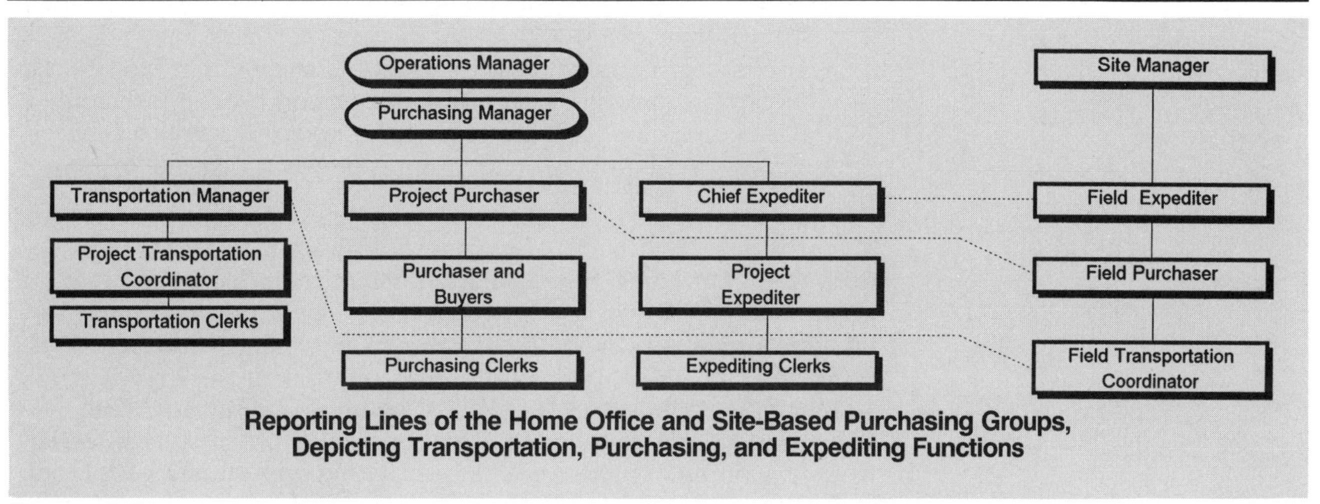

Figure 3.3

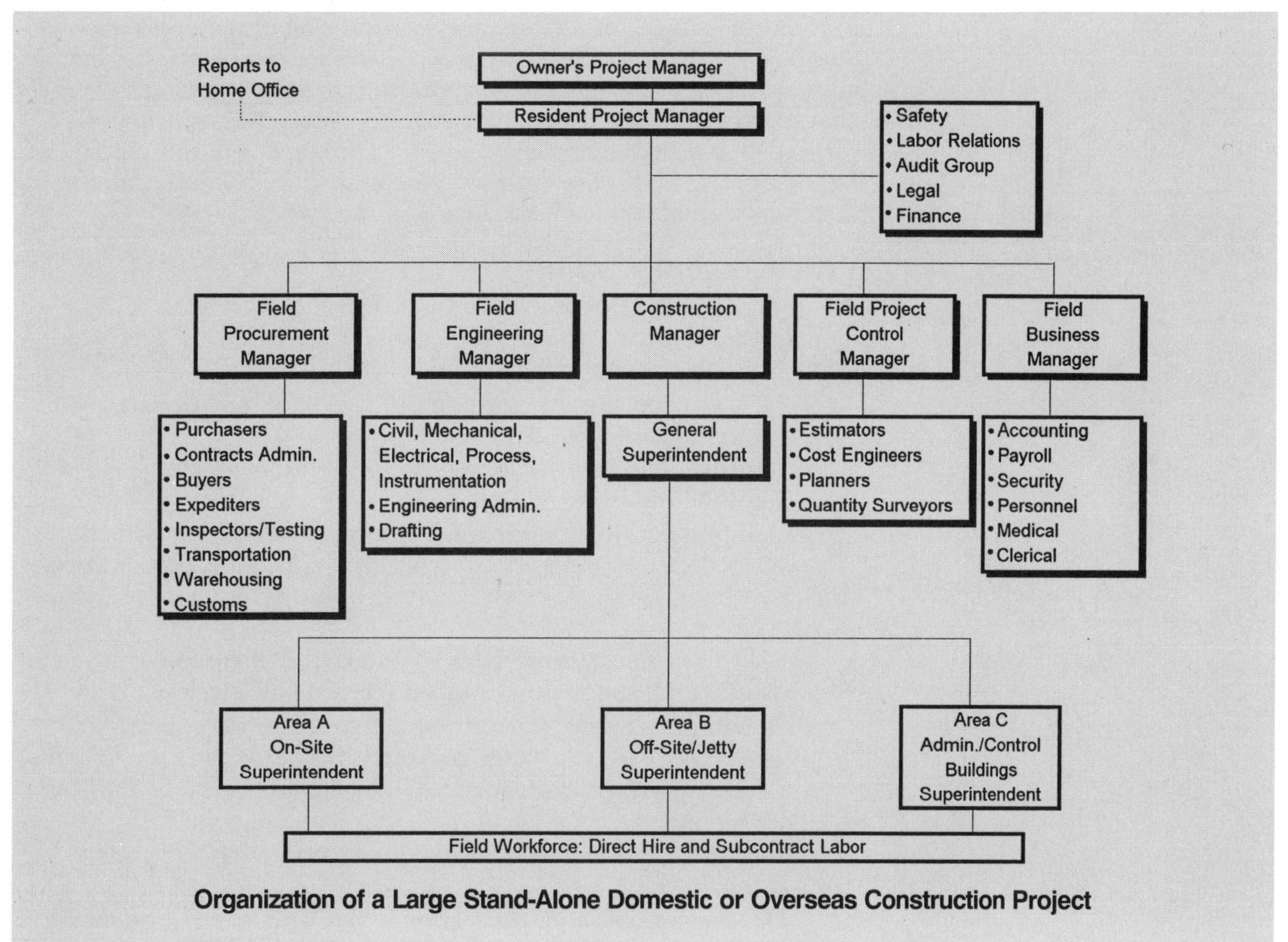

Figure 3.4

- Prevention of "buck passing" by defining individual responsibility for elements of work.
- Use of organizational chart to assist management in the review of expansion/contraction requirements and programs that are the end result of current and future projected workloads.
- Reduction of jurisdictional disputes between different groups.

Duties and Responsibilities

The procurement group's primary responsibility is to assist the project team to perform their responsibilities within the project's pre-established budget and schedule goals. The procurement group provides direct and indirect assistance in obtaining, at the most competitive price, appropriate materials, equipment, and services, and arranging for their safe delivery on or before the contracted date. The group also communicates ideas concerning potential cost savings, work improvement methods, bulk purchasing opportunities, different installation techniques, and alternate sources of supplies.

For each new construction project, the procurement group establishes administration and execution policies and procedures for efficient planning, requisitioning, purchasing, evaluating, negotiating, expediting, source inspecting, and transporting of materials and equipment. The procurement group also enacts policies for the administration and eventual close out of purchase orders and subcontracts for each project. Purchasing within the construction project should be undertaken only in accordance with these policies. In addition, a list of personnel who are authorized to make purchasing decisions should be established early in the project.

Procurement Activities

The procurement responsibilities for a typical construction project include the following tasks, which may be increased or decreased depending on specific project requirements.

- Prepare procedures regarding procurement of materials, equipment, services, and subcontracts.
- Select subcontractors, suppliers, and vendors for prequalification and bidding.
- Maintain data base containing information about qualified vendors, consultants, contractors, and subcontractors (names, addresses, union or nonunion, size of project undertaken, and so on).
- Provide comprehensive background check of potential consultants, contractors, subcontractors, and vendors.
- Obtain from appropriate group a detailed scope of work to ensure complete coverage of owner or contractor requirements.
- Evaluate proposal for compliance with commercial and technical requirements.
- Rate proposals with regard to commercial terms, experience, and capability.
- Prepare contract to ensure compliance with owner or contractor requirements with regard to insurance, secrecy, patents, purchasing, and so on.
- Assist with negotiations in award of purchase order or recovery of claims.
- Manufacture various types of preprinted purchase orders and

contracts that reflect standards and terms that can minimize work effort.
- Assist in the preparation of special conditions for purchases and contracts.
- Ensure that an acceptable degree of competition has been achieved in obtaining bids (obtain at least three bids; more than seven bids can be considered counterproductive).
- Assist in renegotiation of bids in case of changes in scope prior to order or contract placement.
- Develop new sources for equipment and materials, to minimize single sourcing.
- Participate in negotiations of purchased items and contractual matters.
- Inform unsuccessful bidders.
- Assist home or field office locations in seeking out, identifying, and soliciting quotations from woman-owned and minority-owned businesses (WBE and MBE). (This situation applies to publicly funded and some private construction projects.)
- Assist in the preparation and acceptance of purchase orders, letters of intent, engineering and construction (labor only) contracts, subcontracts, and so on.
- Review invitations to bid for adequate scope definition, to stimulate a maximum competitive effort by bidders.
- Arrange for expediting and inspection services.
- Administrate changes to material/equipment purchase orders and subcontracts.
- Coordinate submittal of shop drawings and data requirements.
- Arrange for shipping and transportation.
- Assist in the preparation of backcharges to vendors, suppliers, and subcontracts.
- Assist in the invoice approval cycle.
- Participate in purchase order contract and subcontract close-out activities, such as final reports.

Job Descriptions

The following are typical descriptions of jobs involved in construction purchasing. The examples illustrate the work experience, background, and skills that are required for personnel assigned to a contractor working in civil engineering, manufacturing/industrial, or institutional types of construction. The positions described are typical for purchasing personnel assigned to either the home office or a construction site location for medium to large contractors. Smaller organizations can adapt, modify, and combine some of the job descriptions as appropriate to their specific needs. Additional job descriptions may be developed and incorporated into the purchasing procedure.

Purchasing Manager

Responsibilities

Supervises all purchasing department activities and assignments; establishes department policies, operating procedures, and budgets.

Background Requirements

Requires a total understanding of all aspects of construction purchasing.

Activities/Duties

- Coordinates the purchase of equipment, supplies, and services for maximum value in accordance with requirements of engineering specifications and drawings and quality, delivery, and reliability requirements.
- Supervises, trains, and promotes personnel when appropriate. Establishes salary scales and recommends salary increases.
- Establishes staffing needs for current and future workload.
- Plans and assigns work. Formulates standards of performance and directs subordinates in meeting these standards. Produces appraisals of subordinates' work performance.
- Supervises distribution of bids and inquiries, receipt of proposals, negotiations, commercial bid evaluations.
- Evaluates subcontractors, vendors, and suppliers who might be sources of supply and installation. Consults with engineering and construction personnel to clarify requirements and suitability of available materials and equipment for possible project application.
- Supervises activities of the traffic supervisor and his or her staff for domestic and foreign projects. Arranges for services of freight forwarders and sea and air carriers where applicable.
- Maintains close business relationship with manufacturers, vendors, and sales representatives to develop current and new sources of product and pricing information.
- Supervises activities of all field purchasing personnel.
- Notifies senior management of any conditions or potential trends that may influence the supply of materials and equipment.
- Establishes and supervises maintenance of files, data, and lists of approved vendors, suppliers, contractors, and subcontractors.

Supervisory Relationships

Reports to General Manager of Operations. Supervises all purchasing department personnel at home office, branch office, and construction site locations.

Project Purchaser/Purchasing Agent

Responsibilities

Purchases all materials, equipment, supplies, subcontracts, and services as required for each construction project.

Background Requirements

A thorough knowledge of purchasing methods, techniques, and procedures, gained by actual purchasing experience; skill in human relationships; working knowledge of the construction industry, construction projects, and construction practices. In smaller construction organizations the purchasing agent may perform several additional functions within the organization, such as office administration, project management, and subcontract administration.

Activities/Duties

- Works with the project team to establish purchasing requirements and procedures for the most efficient and economical procurement of materials, equipment, and services.
- Establishes, with the assistance of the planning engineer and the

project master schedule, all long lead times for equipment items to be purchased.

- Evaluates terms and conditions of purchase order for adherence to established standards; drafts, modifies, and recommends any changes in terms and conditions, seeking guidance from lawyers when necessary.
- Communicates with Manager of Purchasing regarding staffing availability, needs, and requirements for project purchasing activities.
- Plans and schedules work for Purchasers and Buyer-Expediters, establishing tasks and priorities to meet target dates with best use of manpower, resources, and budget.
- Attends project team meetings and coordinates purchasing activities with other project team members assigned to the project. Produces minutes and conference notes of all meetings related to purchasing activities.
- Produces and establishes measurable standards for work performance within purchasing department; directs subordinates in meeting these standards.

Supervisory Relationships

Reports directly to Manager of Purchasing. Supervises purchasing personnel under his or her control on assigned projects including purchasers, buyer-expediters, and/or clerical staff.

Purchaser/Assistant Purchasing Agent

Responsibilities

Purchases materials and equipment, and supplies subcontracts and services for the project.

Background Requirements

Requires experience in purchasing methods and procedures; knowledge of construction methods, materials, and equipment; human relations skills for dealing with subcontractors, suppliers, vendors, and project team members.

Activities/Duties

- Coordinates documents related to requests for bids and transmits it to subcontractors, suppliers, and vendors; prepares all necessary documentation for contracts or purchase orders on the basis of engineering and commercial award recommendations.
- Supervises the overall purchasing and purchasing administration function.
- Locates sources of materials, goods, and equipment. Analyzes, reviews, evaluates, and conditions bids where necessary for commercial acceptability of price, terms, and conditions. Confers with Purchasing Agent in selection of subcontractors, suppliers, and vendors.
- Coordinates activities between subcontractors, suppliers, vendors, and project team personnel.
- Conducts research to develop new sources of materials and equipment; recommends new materials, products, or services to replace those currently being purchased and utilized.
- Works with project team members to develop methods of possible cost reduction and cost containment in the purchasing effort,

applying cost-saving techniques such as material substitution, alternate supply sources, and value engineering methods.

Supervisory Relationships

Reports to the Project Purchaser/Purchasing Agent. Assists Project Purchaser/Purchasing Agent in the supervision of personnel assigned to project.

Purchasing Clerk

Responsibilities

Performs clerical functions and activities to assist Project Purchaser/Purchasing Agent, Purchasers, Buyers, and Expediters in the execution of their responsibilities.

Background Requirements

Requires familiarity with clerical methods (filing, letter writing, computer applications) and ability to check invoices submitted by subcontractors, vendors, and suppliers.

Activities/Duties

- Prepares purchasing department reports as required.
- Prepares routine letters to subcontractors, vendors, and suppliers.
- Uses computerized database system to input information in purchasing department's tracking systems.
- Telephones subcontractors, vendors, and suppliers for information purposes.
- Maintains follow-up file on correspondence to subcontractors, suppliers, and vendors.
- Maintains files on assigned projects.
- Maintains backcharge log.
- Maintains shop drawings and vendor data submittal log.
- Assists in reviewing and auditing invoices submitted by subcontractors, suppliers, and vendors.

Supervisory Relationships

Reports to Purchaser/Assistant Purchasing Agent.

Field Expediter

Responsibilities

Expedites materials and equipment from vendors' and fabricators' shops to the construction site location.

Background Requirements

Requires ability to coordinate and communicate with project and construction management teams, engineering departments, suppliers, and vendors to ensure that equipment items are produced according to schedule. Detailed understanding of fabrication shop procedures, scheduling, and routing is required. Familiarity with transportation practices is desirable to maintain transportation and delivery schedules.

Activities/Duties

- Interprets and explains engineering drawings, fabrication details, shop drawings, specifications, and code requirements.
- Remains aware of layout and configuration, management and organization, work methods, procedures, and standard operating practices of fabrication shops.

- Gathers and maintains relevant and accurate data for ongoing and historical equipment information records. Presents this information in oral and written form to the various project team members, obtaining all necessary shop fabrication reports, manpower schedules, and other necessary supporting documentation. Collects all purchasing site receiving and logging files.
- Attends all site progress meetings and reports on progress or potential problems related to materials and equipment.

Supervisory Relationships

Reports administratively to the Chief Expediter in home office and to Site Construction Manager at site location. Supervises Assistant Expediters.

The Purchasing Plan

When a new construction project commences, a purchasing plan is usually initiated. The purchasing plan must consider materials, equipment, and subcontracts.

The key components of two types of purchasing plans—for subcontracts, and for materials and equipment—are listed below. The two plans are similar in many ways; the basic difference between them is the labor element associated with subcontracts. Coordination with other subcontractors, insurance requirements, and payment terms require careful planning when formulating a subcontract.

Purchasing Plan for Subcontracts

- Receipt of engineering data, specifications, and drawings by purchasing department.
- Preparation of list of approved subcontractors.
- Production of subcontract packages with terms and conditions including hourly rates and unit rates (material and labor) for additional work; transmittal to bidders.
- Arrangements of clarification meetings if required.
- Receipt of bids; performance of technical and commercial evaluation.
- Condition of bids for missing scope items or exclusions.
- Recommendation and award of subcontract.
- Issuance of final subcontract for signature.

Purchasing Plan for Materials (Commodities or Bulk) and Equipment

- Receipt of engineering data, specifications, and drawings by purchasing department.
- Preparation of list of approved vendors and suppliers.
- Production of inquiry package with terms and conditions; transmittal to bidders.
- Arrangement of clarification meetings if required.
- Receipt of bids; performance of technical and commercial evaluation.
- Condition of bids for missing scope terms or exclusions.
- Recommendation and award of contract.
- Issuance of purchase order.

Figure 3.5 is an example of an equipment purchasing plan. Similar plans can be used for materials and subcontracts.

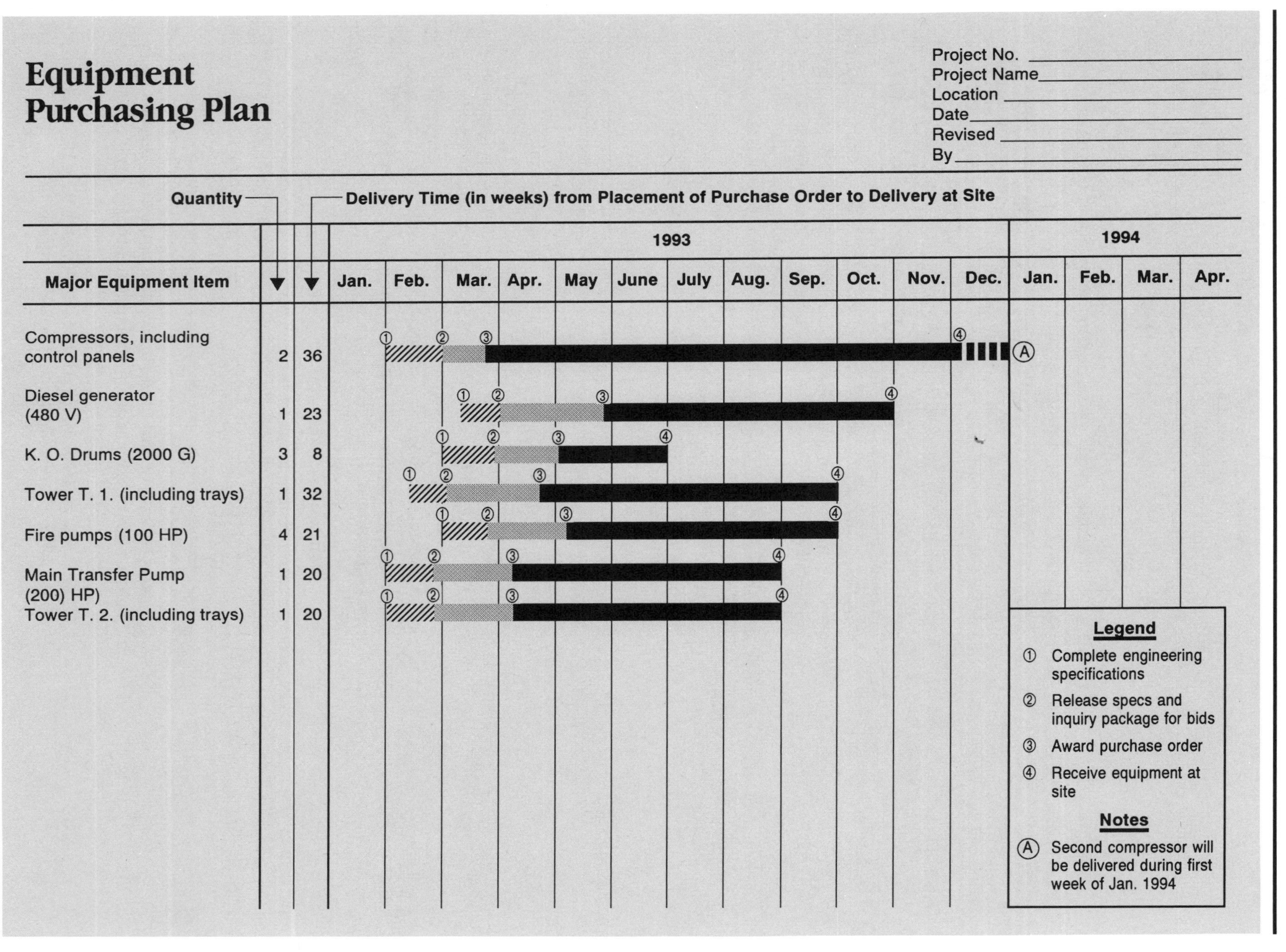

Equipment Purchasing Plan
Project No.
Project Name
Location
Date
Revised
By
Quantity
Delivery Time (in weeks) from Placement of Purchase Order to Delivery at Site
1993
1994
Major Equipment Item
Jan.
Feb.
Mar.
Apr.
May
June
July
Aug.
Sep.
Oct.
Nov.
Dec.
Jan.
Feb.
Mar.
Apr.
Compressors, including control panels
2
36
Diesel generator (480 V)
1
23
K. O. Drums (2000 G)
3
8
Tower T. 1. (including trays)
1
32
Fire pumps (100 HP)
4
21
Main Transfer Pump (200) HP)
1
20
Tower T. 2. (including trays)
1
20
Legend
① Complete engineering specifications
② Release specs and inquiry package for bids
③ Award purchase order
④ Receive equipment at site
Notes
Ⓐ Second compressor will be delivered during first week of Jan. 1994

Figure 3.5

On many construction projects the owner may choose to buy certain items. These items typically consist of furniture, fixtures, and specialist equipment (the kind of materials and equipment covered in Division 10, Specialties; Division 11, Equipment; Division 12, Furnishings; and Division 13, Special Construction of the CSI MasterFormat). The general contractor's responsibility usually is to receive, unpack, and install these items. This process must be planned; a scope of work must be developed regarding the exact items and materials that the owner will supply.

The Purchasing Flow Chart

The purchasing cycle begins when the purchasing department receives a scope of work statement or a properly executed materials/equipment requisition. The scope of work or requisition is usually supplied by the engineering group; it consists of drawings, specifications, and any special performance requirements. The flow chart shown in Figure 3.6 illustrates the sequence of steps leading from the requisition to the final executed purchase order, contract, or subcontract. The bidding process is discussed in detail in Chapter 5 (Contracting, Subcontracting, and Administration).

Purchase Orders

A *purchase order* contains the results and conclusions of the purchasing process in a contractual format, describing the articles being purchased, the price of those articles, and the method of delivery. A purchase order becomes a contract between buyer and seller only when it is formally accepted by responsible representatives of both organizations.

Once the vendor or supplier has been selected and all the required approvals have been obtained, it is very important to produce and issue the purchase order as soon as possible. Early issuance of the purchase order will prevent any possible delays to the project. It is good practice to get the signed acceptance copy of the purchase order from the vendor or supplier within five working days.

Purchase orders must be worded carefully: all essential elements of the transaction must be handled in a manner that will not cause future misunderstandings or problems. In this way, the necessity for lengthy post-award correspondence and change orders can be minimized.

Figure 3.7 is an example of a first page or cover sheet for a purchase order. This example can be modified or adapted for future specific purchasing activities.

Figure 3.8 is an example of a purchase order continuation sheet, which can be used in addition to the purchase order.

Language and Elements of the Purchase Order

Every construction-related organization must develop its own standard terms, clauses, conditions, and language for purchase orders. Prior to creating specific language for the purchase order, a clause to protect the buyer should be developed. Following is an example of such a disclaimer.

> *This purchase order contains all terms and conditions of the parties' agreement concerning the purchase of materials, equipment, and services. It may not be added to, modified, or superseded except in writing and approved by an authorized representative of the buyer. Different or additional terms, clauses, and/or conditions in*

seller's responses will be rejected; no subsequent conduct by the buyer shall be deemed to be an acceptance of these additional terms and conditions.

This article or clause should serve to protect the buying organization from possible future conflicts.

It is good practice to list the various standard terms, clauses, conditions, and language and to include them all on the reverse side of the purchase order document depicted in Figure 3.7. The following is a list of these standard items that should be considered and expanded upon on the reverse side of the purchase order.

- Changes or additions to the purchase order
- Bid, payment, or performance bonds
- Delays in work
- Conflicts and stop work orders

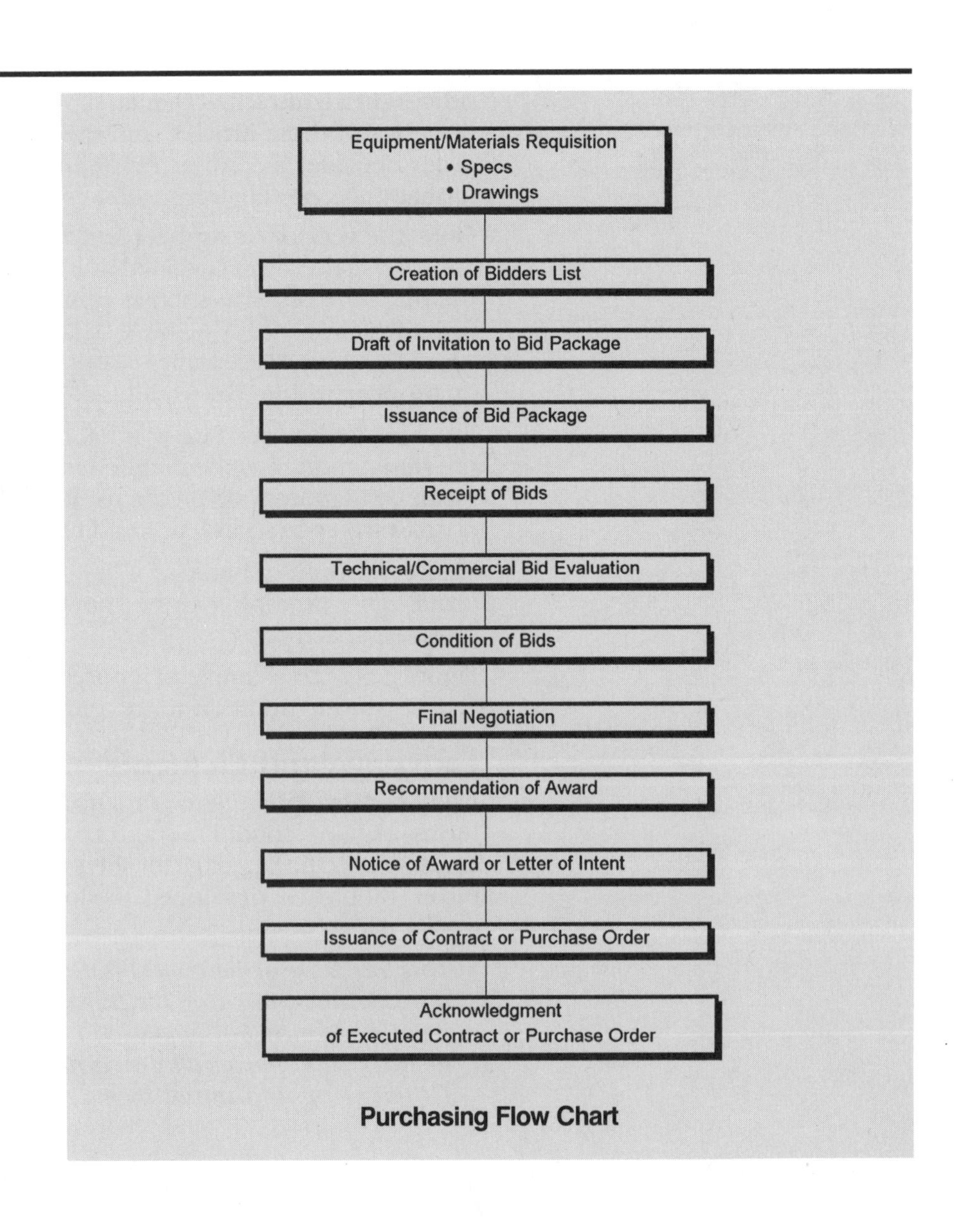

Purchasing Flow Chart

Figure 3.6

Purchase Order

(This is an offer to purchase on the terms and conditions stated on the reverse side of this document)

Project No. ____
Project Name ____
Project Location ____
Date ____
Page 1 of ____
Prepared By ____

Purchase Order No.

To: ____

Ship To: ____

Item No.	Quantity	Description	Code or Equip. No.	Unit Price	Extension Value

Confirming Order Placed With

Total Value **$**

Cost Code

Payment Terms

Invoice Instructions
Send ☐ (____) copies of invoice with original bill of lading to:

Shipper

☐ Sales Tax Included

Ship Via

Weight/Lbs.

Cube/CF

Shipping Terms

Promised Shipment

The Attachments Below Are Hereby Made a Part of This Contract

☐ General Terms and Conditions ____

☐ Drawings ____

☐ Specifications ____

☐ Vendor's Proposal ____

☐ Other ____

Approval By

Title

Company

Date

Seller hereby accepts and agrees to all terms and conditions, and only those terms and conditions contained on the reverse side of this purchase order.

Accepted by

Title

Company

Date

Distribution ____

Action ____

Figure 3.7

Purchase Order Continuation Sheet

Project No. ____________________
Project Name ____________________
Project Location ____________________
Date ____________________
Page ______ of ____________________
Prepared By ____________________
Revision No. ____________________

Purchase Order No.

Item	Quantity	General Description of Materials/Services	Unit Price	Extension
			Total	$

Figure 3.8

- Lien indemnifications
- Force majeure provisions
- Drawing and specification reviews
- Indemnity
- Patent indemnity
- Termination for default
- Termination for convenience of buyer
- Inspection of the work
- Expediting
- Delivery of equipment and materials
- Payment terms
- Warranties and guarantees
- Assignments and subcontracting
- Secrecy and confidential information
- Buyers' property
- Insurance requirements
- Compliance with laws
- Fair Labor Standards Act, OSHA, Equal Employment Opportunity
- Variations in quantity
- Arbitration

Purchase Order Log

Every action connected with the placement and execution of a purchase order and any subsequent modifications or amendments should be incorporated into the purchase order log (sometimes referred to as the purchase order register). This document is maintained as part of the purchasing department's permanent records related to a particular project. An example of a purchase order log is shown in Figure 3.9.

Change Orders

A change order is a legally enforceable addition to or deletion from a purchase order or contract. The change order should be written in a clear and concise manner by the purchasing department. Figure 3.10 is an example of a change order related to a purchase order. (An example of a contract change order is contained in Appendix A.)

Change orders to equipment and materials purchase orders usually occur when there is a modification in the engineering and design requirements. These changes typically are transmitted to the supplier or vendor in the form of a bulletin.

The purchasing department will transmit a written request to the subcontractor, supplier, or vendor to submit a price for the proposed changes or modifications before being incorporated into the purchase order. Upon receipt of the price, the purchasing department will compare it with an estimate of the change produced by an in-house estimator who is familiar with the specific work being modified.

Dealing With Change Orders: The following options are available when dealing with a change order or modification.

- Accept the price if it is consistent with the in-house estimate; advise the supplier or vendor to proceed with the work.
- Decline the price and, if possible, negotiate a more favorable cost with the vendor or supplier.
- Reconcile, modify, or minimize the design work to reduce the

Purchase Order Log

Project No. ____________
Project Name ____________
Report No. ____________
Date ____________
Page ______ of ______

P.O. Number	P.O. Date	Revision No.	Vendor/Supplier Name	Description	$ Value	Delivery Date

Comments: ____________

Figure 3.9

Purchase Order Change Order

Project No. ______________
Project Name______________
Purchase Order No. ______________
Change Order No. ______________
Page ______ of ______________
Date______________

To:

Attn:

Telephone No.

Fax No.

Confirmed to:

Date

Details of Change Order:

Item No.	Quantity	General Description of Materials/Services	$ Extension
		Total Value of Change Order	$

Original purchase order amount $ ______________

This change order value $ ______________

Revised purchase order amount $ ______________

Effects delivery by __________ days (+ or –)
(add or subtract as necessary)

Soder Construction and Development Corp.
Approval Authority

By

Title Date

Seller hereby accepts and agrees to all terms and conditions of Purchase Order No. ______________

Dated __________ __________ and this Change Order.

Accepted by Title

Company Date

Figure 3.10

cost impact of the change or modification.
- Ascertain from the engineering and project management groups whether the design changes or modifications can be excluded entirely.

Purchase Order Files

The volume of correspondence, drawings, and paperwork that accumulates during the inquiry, negotiations, and eventual placement of each purchase order or subcontract makes it necessary for files to be arranged and maintained in a neat and orderly manner.

A six-part file system is helpful in keeping the various documents related to the purchase order in a convenient format. The file system might be organized in the following manner.

- Part 1. Signed purchase order, any change orders to original purchase order, and any subsequent backcharges or claims. Copies of invoices and payment records or details.
- Part 2. Specifications and any subsequent specification modifications.
- Part 3. Original bid drawings and any subsequent drawing revisions.
- Part 4. Correspondence to and from supplier, vendor, or subcontractor.
- Part 5. Expediting reports, delivery documents, shortage reports, damage reports, inspection reports.
- Part 6. Correspondence prior to placement of purchase order, copy of bid tabulation summary, recommendation details.

Operating Principles and Ethics

Successful purchasing depends on the performance of purchasing activities based on certain basic principles, which have been discussed in this chapter. The utmost care must be taken when conducting purchasing business activities. The purchasing department's relationship with suppliers, vendors, and subcontractors projects the image of the contracting organization; these relationships can either enhance or injure the company's reputation. Therefore, the purchasing department should conduct all business dealings with suppliers, vendors, and subcontractors in a fair and courteous manner.

The purchasing department should provide the supplier, vendor, or subcontractor with clear, concise instructions regarding project requirements. No advantage should be made of errors contained in bids or proposals.

Actual performance of suppliers, vendors, and subcontractors should be closely monitored by the purchasing department. Information regarding unsatisfactory performance should be the basis for exclusion from future projects.

The relationship between the purchasing department and outside suppliers or workers is not the only relationship that merits consideration. The purchasing group should communicate and coordinate with all project team members in a positive manner. The planning and execution of purchasing activities has a significant impact on the overall success of the construction project.

Undesirable business practices such as "bid shopping" should be avoided at all times. Bid shopping enables the general contractor to increase his or her potential profit at the expense of vendors, suppliers, or subcontractors. When the contractor has been awarded a construction contract, he or she might shop around, letting subcontractors and suppliers know the prices included in the bid. The expectation is that lower bids can then be obtained. This practice has caused much concern and debate within the construction industry.

Finally, the Purchasing Manager should be progressive, staying abreast of the latest trends. The purchasing management and staff should maintain an open mind regarding new methods, techniques, procedures, and materials that might positively influence the operation of the department and the future success of the company as a whole.

In conclusion, the following guidelines should serve as basic ethics of professional conduct for the purchasing department.

- Consider the interests of the company and the client in all business transactions.
- Offer all bidders opportunities to bid and compete on equal terms and conditions.
- The bids of competing subcontractors and vendors should not be made known before the award of the subcontract and purchase order, nor should they be used by the general contractor to obtain lower bids.
- Strive for honesty and integrity in all purchasing and selling activities.
- When a general contractor is informed that he or she is the successful bidder, the general contractor should award subcontracts and purchase orders to the parties whose bids he or she utilized in compiling the successful bid.
- Purchase without prejudice or favor, seeking to obtain the maximum value in the form of price, quality, service, reliability and delivery.
- Consider confidential all information obtained during the bidding cycle and contract negotiations that concerns bid prices, terms, and conditions of purchase and performance.
- Treat in a fair and consistent manner all project team members, colleagues, and coworkers regardless of race, sex, age, religion, or national origin.
- Never seek or accept gifts or favors from suppliers, vendors, or subcontractors that may be offered because of a business relationship.

Chapter 4

Quality, Inspection, Expediting, and Transportation Activities

Chapter 4

Quality, Inspection, Expediting, and Transportation Activities

The activities related to quality, inspection, expediting, and transportation for any construction project are specialized elements of the overall purchasing management program. Research and experience have indicated that most successful construction projects have largely depended on detailed front-end planning and scheduling related to these four important activities.

Overview of Quality Assurance and Quality Control

The quality of construction has always been of major interest and concern to owners and end users of projects or facilities. The quality of construction materials, equipment, products, and installation can affect the cost, schedule, and eventual operating life of a building or facility. It is important to fully understand the terms *quality, quality assurance,* and *quality control* as they relate to construction.

Quality refers to the excellence, merit, or fineness of a finished product. Construction materials and equipment can be considered of good quality when they fully conform to the drawings, specifications, and requirements of the specifying architect or engineer.

Quality assurance refers to the presumption and verification that suppliers, vendors, and sub-vendors have conformed to a procedure that was established prior to the fabrication and manufacturing process. (Quality assurance is often abbreviated as *QA.*)

Quality control includes ensuring that the installation of materials and equipment is performed in compliance with particular specifications, drawings, and installation procedures. Quality control also includes the actual testing, inspection, and documentation for materials and methods used in the installation process. (Quality control is often abbreviated as *QC.*)

Owners, engineers, architects, contractors and subcontractors should communicate on a regular basis with their employees, vendors, and suppliers regarding the need for quality materials and equipment. Materials and equipment of poor or substandard quality will eventually cause problems in a building or facility; there will eventually be a breakdown or failure. The best time to address this potential problem is prior to the design and installation phase of the project.

Careful planning and communication related to quality considerations in the early stages of a project will ensure that all project team members are cognizant of the owner's quality requirements. This understanding will dilute or solve any problems that may occur during the installation phase of the construction project.

Quality Assurance/Quality Control Flow Chart

The flow chart in Figure 4.1 outlines the steps that are required for a typical quality assurance/quality control program.

Quality Assurance/Quality Control Checklist

The following is a checklist of questions that can be addressed in determining the quality needs of a construction project.

- Which organization is responsible for determining the quality considerations and requirements for the project – owner, architect, engineer, or contractor?
- Do specifications and drawings address the owner's quality needs?
- Does the contractor's purchasing department fully understand the quality requirements of the construction project?
- Does the contractor's prequalification and selection process for suppliers, vendors, and subcontractors include quality

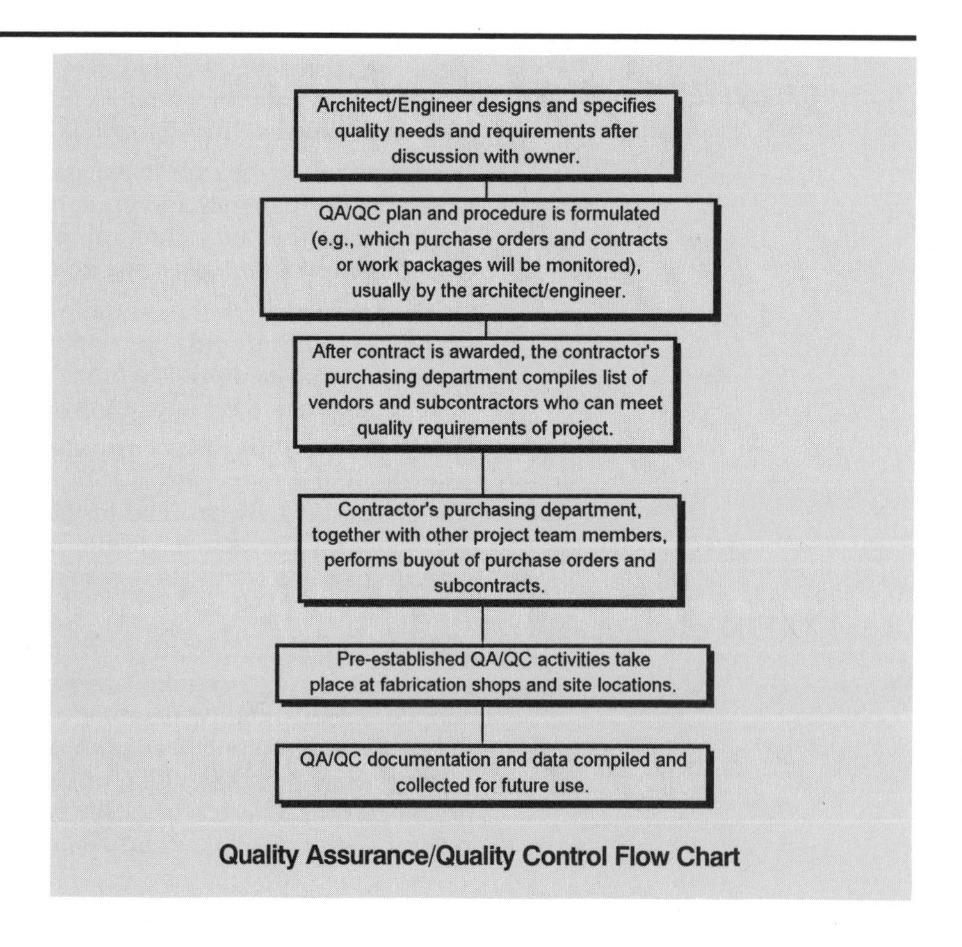

Quality Assurance/Quality Control Flow Chart

Figure 4.1

considerations? Are the records of past performance of bidding organizations verified?

- Is quality a major consideration in the process of evaluation and selection of suppliers, vendors, and subcontractors?
- Do suppliers, vendors, and subcontractors have a pre-established QA/QC program? If so, do they use this program for every project?
- Are suppliers, vendors, and subcontractors aware of the project's quality acceptance methods?
- Are suppliers, vendors, and subcontractors aware of product and data submittal requirements?
- Is full-time, on-site quality control required at the fabrication or manufacturing locations, or are periodic visits and audits sufficient?
- Which organization (architect/engineer or contractor) will coordinate the quality inspection visits to the supplier's, vendor's, or subcontractor's workplace?
- Does the construction project have the qualified personnel and resources to confirm that specified materials and equipment are being properly incorporated into the construction project?
- Which organization is responsible for specification verification, acceptance, storage, and eventual installation of materials or equipment?
- Are materials and equipment checked for completeness, damage, or defects prior to delivery or receipt at the job site?
- Is a QA/QC program being used at the construction site?
- Are suppliers, vendors, and subcontractors notified of damage or defects to materials and equipment?
- Is a procedure in place for suppliers, vendors, and subcontractors to repair and rectify damaged or defective materials and equipment?

Shop Drawings, Samples, and Product Data

The approval of shop drawings, samples, and product data is a basic and important element of the overall QA/QC program. There are various types of shop drawings; they are created and produced by contractors, subcontractors, vendors, suppliers, and equipment manufacturers. They illustrate and describe construction materials, functions, fabrication techniques, dimensions, installation sequences, and other relevant information for the incorporation of specific elements or items of work into the completed construction project. Samples and product data are additional sources of this type of information.

The following types of drawings and categories of data are collectively considered shop drawings, samples, and product data.

- Manufacturing sequence drawings
- Fabrication and detail drawings
- Final position drawings/location diagrams
- Final and temporary erection sequence drawings
- Manufacturer's standard general arrangement drawings
- Catalog cuts and performance data
- Wiring and control arrangement drawings
- Design data sheets
- Samples and examples of finished product
- Coordination and interface drawings and sketches

- Installation and lifting arrangement drawings

When the general contractor receives shop drawings, samples, and product data from the vendor, supplier, or subcontractor, he or she is responsible for reviewing and checking their content and validity against the construction project's contract drawings and specifications. The general contractor will then send them on to the architect or engineer for further review and subsequent approval. If the architect or engineer agrees with the shop drawings, samples, and product data, he or she will approve them and return them to the general contractor, who will then return them to the relevant vendor, supplier, or subcontractor. However, if the architect or engineer finds errors, he or she will return the data to the general contractor with an explanation for the rejection. The general contractor will in turn relay this information to the vendor, supplier, or subcontractor for corrective action.

It is important that the review and approval of shop drawings, samples, and product data be performed in a timely manner. If the review and approval time are delayed, the delivery of materials and equipment to the project site will be delayed. The flow chart in Figure 4.2 depicts a typical review and approval cycle for shop drawings, samples, and product data.

Shop Drawings, Samples, and Product Data Register

The planning, scheduling, and expediting of shop drawings, samples, and product data and the subsequent review and approval of this data are critical activities that relate directly to the overall success of any construction project. Because of the important nature of these activities, the purchasing department should maintain a project register or log for all shop drawings and related data. This register should indicate:

- The project number, name, and location
- The applicable purchase order or subcontract number
- The required listing of submittals for each purchase order or subcontract
- The time frame or specific end date for issuance of submittals to general contractor
- The status of submittals in the review and approval cycle, i.e.,
  - Being reviewed by general contractor
  - Being reviewed by architect/engineer
  - Approved and return to vendor, supplier, or subcontractor for final fabrication or construction
  - Modifications or changes that need to be completed prior to resubmittal of data for approval

Inspection

Because a construction project must be built according to the quality and statutory requirements of the owner or end user, and must also comply with local building codes and all applicable statutory requirements, *inspection* is the activity that ensures compliance is achieved. Inspection ensures compliance with such requirements as:

- Contract drawings and specifications
- Purchase order terms and conditions
- Subcontract agreements and obligations

- Shop drawings, samples, and related product data
- Statutory regulations such as OSHA (Occupational Safety & Health Act); government agencies; city building inspectors; fire marshals; and state elevator inspectors

Planning for Inspection

Front-end planning for inspection should take place at an early stage in the project, preferably during the design phase. The inspection plan should remain in effect until the completion of the construction effort. A determination should be made during the engineering and procurement phases as to which materials, equipment, and installation elements are to be inspected. Fabrication shop inspection, for instance, is usually limited to engineered equipment such as boilers, compressors, distillation towers, elevators, and certain key fabrication components such as piping, structural steel, and precast concrete elements.

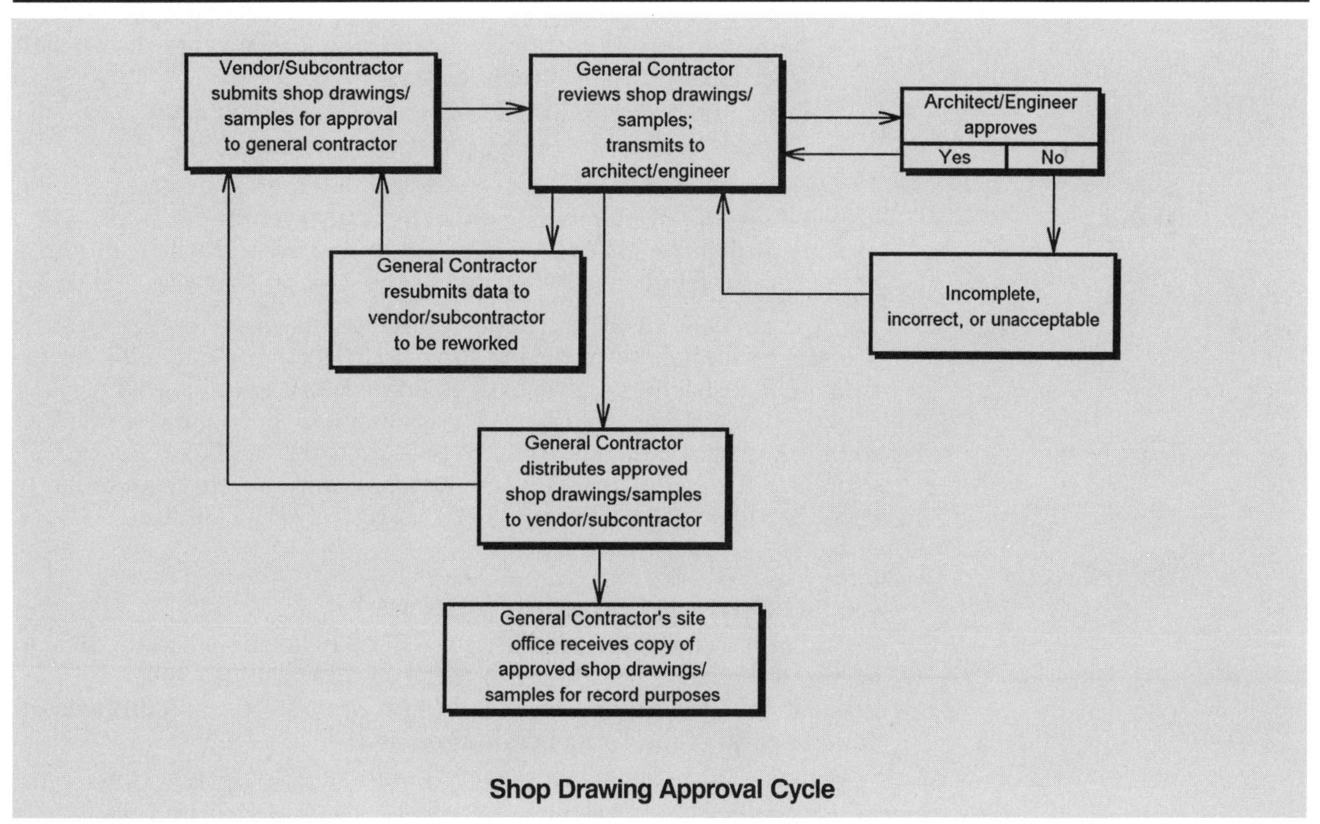

Figure 4.2

The inspection plan must recognize the various inspection activities that will be performed either at a specific time or on a continuous basis. These inspection activities might consist of the following:

- Review of the effectiveness of the fabricators', vendors', suppliers', and subcontractors' quality control methods and techniques.
- Examination of destructive and nondestructive testing methods and procedures. Some of the most frequently used tests are concrete strength, soil bearing, pile testing, and structural steel testing. ASTM (American Society of Testing and Materials), UL (Underwriters Laboratories), and ASME (American Society of Mechanical Engineers) are the organizations that specify the testing parameters of construction materials. The architect or engineer often contracts with an independent testing agency or consultant to perform some or all of these tests and inspection activities. The reason for this is that a general contractor could influence an independent testing firm if the general contractor is paying and directing the activities of that firm.
- Mechanical or electrical testing, including performance and operation testing.
- Review and monitoring of all documentation, test results, and other pertinent data and records.
- Check of dimensions, sizes, and thicknesses against contract drawings and specifications.

The inspection plan should also consider the coordination and communication of inspection personnel, including:

- Notification of the vendor, fabricator, or subcontractor of the required test dates and/or acceptance tests.
- Use of certain forms and certificates in the inspection process.
- Name and telephone number of individual at fabrication shop with whom inspectors will be in contact.
- Entry and access information related to gaining entry into fabrication shop, including the use of office facilities.

The inspection plan should also consider the requirements for qualified inspection staff, the number of individuals involved in the inspection process, and the length of time that these staff members will be required at the various fabrication shops and at the construction site.

The architect-engineer or general contractor may sometimes require specialized inspection or testing services. This might be in addition to daily field inspection and usually is limited to a specific area of expertise. For example, an outside consultant might be hired to conduct soil boring tests, concrete strength tests, or x-ray welding tests. In general, consultants may be used for inspection of items requiring specialized knowledge or expertise that typical site inspectors do not possess.

Inspection by Outside Agencies

In addition to the inspection activities already detailed, certain outside agencies such as OSHA, fire marshals, city and state building departments, and so on, may need to be considered in the inspection and approval of the construction process.

Fabrication shop inspection and field inspection are two key elements in the quality control/quality assurance program. Inspection plans should be targeted at preventing and minimizing defects and deficiencies. If defects and deficiencies do occur, the overall inspection plan should ensure that these problems are found and corrected early in the construction process. Costly rework and possible delays in construction completion can be avoided in this way.

Fabrication Shop Inspection

Fabrication shop inspection is often referred to as *quality surveillance* or *quality inspection*. Fabrication shop inspection includes all the quality surveillance activities performed at a vendor's, supplier's, or subcontractor's fabrication shop or manufacturing facility. These inspections consist of examining and reviewing the fabrication process as well as all testing routines and requirements that are part of the contractual agreement between the buyer and seller.

The main inspection activities performed in the fabrication shop include those listed under the "Planning for Inspection" section of this chapter. Fabrication shop inspection is basically a preventative measure; when performed correctly and diligently, it can be a key element in ensuring that a construction project meets its quality requirements and goals.

Field Inspection

Field inspection is just as important as fabrication shop inspection. The field inspection staff makes a thorough examination of all delivered materials and equipment for both quality and quantity. This inspection should be performed during the processes of unloading, erection, and, if necessary, storage. A thorough examination (including review of any delivery notes or transport documentation) of materials and equipment during or shortly after receipt will determine whether they reflect and meet the requirements of the contract drawings and specifications. This inspection also serves to locate any defects, damage, or missing items of material or equipment.

In general, field inspection activities ensure that the installing subcontractor's work is being performed in accordance with the requirements of the contract documents (i.e., drawings and specifications). Inspection for substantial completion and the final inspection for acceptance and eventual delivery, or "hand-over," to the client are also part of the field inspection cycle.

The architect-engineer or general contractor assigns qualified field inspectors to the construction site to safeguard against poor workmanship, defects and deficiencies. Work that does not conform to the contract drawings and specifications will be rejected and instructions for correction will be issued to the responsible installing contractor. Records and details should always be maintained on all unacceptable material, equipment, and installation work.

Inspection Checklist

The following is a checklist of questions that should be addressed prior to the commencement of any inspection activities.

- What magnitude and level of inspection will be necessary on the construction project?
- Will in-house or contracted inspection services be utilized? What are the costs of these alternatives?
- What inspection procedures and reporting formats will be required?

- What tests and construction techniques will have to be witnessed?
- What certifications and documentation requirements will be needed on the construction project? Will the owner require this documentation?
- Is there a procedure for dealing with deficiencies or nonconformance with specifications?
- Is a backcharge procedure in place for deducting money for damaged or defective materials and equipment?
- Is adequate off-loading equipment (i.e., cranes, hoists, forklifts) available at the construction site to safely unload materials and equipment?

Fabrication/Inspection Status Reports

Figures 4.3a and 4.3b are examples of fabrication/inspection status reports. They are specifically designed for the fabrication of valves and structural steel. The purpose of these two inspection reports is to monitor and audit inspection progress and delivery of specific materials, components and equipment. These reports are usually compiled by the architect or engineer's inspection staff, or by the contractor's inspectors. These two forms can be modified and adapted for other inspection applications.

Expediting

As discussed in Chapter 2 (The Overall Procurement Plan), the objective and main function of expediting activities are to ensure that equipment and materials are delivered to the construction project in accordance with the stipulated delivery times described within the purchase order or subcontract agreement.

The timely delivery (in accordance with the project schedule) of equipment and materials is vital to the success of all construction projects. If equipment and materials are not on the construction project when required, the owner, contractor, and/or subcontractor may encounter serious problems, including:

- Delay in construction project completion date
- Increase in costs of equipment and materials
- Additional finance and interest charges
- Productivity loss in the construction implementation phase
- Additional costs related to maintaining the site establishment (trailer, site staff salaries, expenses, etc.)
- Claims by installation subcontractors
- Loss of potential profit

In today's business environment, it is essential that an expediting procedure be developed that can impose an optimal system of pre-planning and subsequent control on the construction procurement process and buy-out activities. The expediting function affects the project from the date of purchase order or contract award to the delivery of fabricated or manufactured equipment and materials to the construction site.

Fabrication/Inspection Status Report/ Valves

Valves 12" and above

Vendor/Supplier Name ABC Valve Co.

P.O. No 030

Sub-Vendor/Supplier Name N/A

Location N/A

Contact N/A

Project No. C125

Report No. 2

Date 10/31/9X

Page 1 of 1

Contract No. N/A

Inspected By J.T.

Legend

C=Complete S=Started F=Fabrication in Progress N=Not Yet Started D=Delivered to Site

Valve No.	Body Castings	Internals	Operators	Machining	Assembly	Testing	Delivered to Site	Remarks/Comments
V101 A	C	C	C	C	C	C	D	Received 10/20/9X. Good Condition
V101 B	C	C	C	C	C	C	D	Received 10/25/9X. 1 Bolt/Nut Missing.
V102	F(100%)	F(100%)	F(50%)	F(25%)	N	N	N	To be shipped 11/21/9X.
V103	F(25%)	F(25%)	N	N	N	N	N	To be shipped 11/21/9X.
V104	F(10%)	N	N	N	N	N	N	To be shipped 12/5/9X.
V105	S	S	N	N	N	N	N	To be shipped 12/5/9X.

Figure 4.3a

Fabrication/Inspection Status Report/ Structural Steel

Vendor/Supplier Name TPM Steel, Inc.

P.O. No. N/A
Sub-Vendor/Supplier Name N/A
Location N/A
Contact N/A

Project No. C 124
Report No. 3
Date 12/21/9X
Page 1 of 1
Contract No. C-003
Inspected by J.T.

Legend

C=Complete S=Started F=Fabrication in Progress N=Not Yet Started D=Delivered to Site

Description	Shop Drawings Approved for Fabrication	Materials Delivered to Site	Fabrication Work	Welding Work	Priming & Painting	Anchor Bolts	Delivered to Site	Remarks/Comments
Mfg. Plant 750 Ton	C	C	C	C	C	C	D	Shop primed paint needs touch up work; was damaged in transit. Received 11/27/9X.
Office Building A 125 Ton	C	C	C	C	S(50%)	S(10%)	N	
Office Building B 150 Ton	C	C	F(25%)	N	N	N	N	
Pipe Racks A.I. 300 Ton	C	S (50%)	F(10%)	N	N	N	N	
Sub Station 76 Ton	C	S (50%)	N	N	N	N	N	

Figure 4.3b

Expediting Methods

There are several types of expediting methods used in the construction industry. Each of these methods has a different level of intensity or magnitude and subsequent cost implications. The three levels of expediting methods used most frequently in the construction industry are discussed in the following sections.

Detailed or Intensive Expediting: Detailed expediting is the most comprehensive and labor-intensive method utilized, and usually is the most productive in maintaining or improving scheduled delivery of materials and equipment. Immediately after the purchase order or subcontract award, the expediting group makes contact with the vendor, supplier, or subcontractor. On smaller projects, this activity might be performed by the project manager or scheduler/planner assigned to the project. The responsibilities of the expediting staff in this approach are listed below:

- Makes contact with vendor, supplier, or subcontractor early in the manufacture or fabrication sequence, and maintains these contacts throughout the fabrication and delivery process.
- Visits vendor's or supplier's fabrication shop to monitor and audit the manufacture and fabrication process.
- Ensures that engineering specifications, cut sheets, and shop drawings are transmitted from the vendor or supplier to the design or engineering group for approvals.
- Monitors and audits vendor or supplier purchase orders with sub-vendors and sub-suppliers and receipt of any subcomponents.
- Remains aware of other projects and orders that have any impact or effect on the progress of the vendor or supplier.
- Coordinates and informs the various project team members of any critical inspection requirements and current delivery date.

A thorough understanding of these activities together with all the minor details, elements, and requirements of the purchase order or contract will enable the expediter to address and work around any potential problems during the fabrication or manufacturing process.

Regular or Planned Expediting: Regular expediting is not as comprehensive or sophisticated as detailed expediting. As the promised delivery date approaches, the expediter may visit the fabrication shop to confirm compliance with the delivery date. Telephone calls and weekly fax-type status reports from the vendor, supplier, or subcontractor may suffice in this level of expediting. In many cases, the expediter will visit the fabrication shop only when a problem or delivery delay has occurred or is about to occur.

In general, regular expediting focuses mainly on data collection and monitoring as opposed to the intensive day-to-day problem-solving actions of detailed or intensive expediting.

Status Report Expediting: As its name implies, status report expediting is less stringent than the two previously described methods. Weekly or monthly telephone contact is maintained with the vendor, supplier, or subcontractor. The expediter might ask the vendor or supplier to issue a written status report that indicates physical progress related to predetermined milestone events. This type of expediting has limited success in preventing potential problems and delays in fabrication or delivery, but the low costs associated with this method make it a cost-effective approach. The vast majority of construction projects utilize this method of expediting.

Expediting Organization

The organization and structure of the expediting function is detailed in Chapter 3 (Purchasing). The group consists of home office expediters and field- or construction site-based expediters. Both home office and field expediters report to the chief expediter. In addition, field expediters interface and report daily to the project site manager. On smaller construction projects, this function can be performed either by the project manager, site manager, planner, or site administrator.

Expediting of Vendor Drawings and Related Items

Purchase orders and subcontracts usually give precise dates for the submittal of vendor drawings, equipment data, and other contractual information that may be necessary prior to the commencement of design, fabrication, or manufacture of construction work. A standard purchase order or subcontract should stipulate submittal dates for the following major items:

- Overall size and weight, including loaded weight of equipment or component.
- Dimensions and footprint size of equipment; location and sizes of any piping connections and nozzles.
- Section details indicating clearance height requirements, foundation sizes and loadings, and anchor bolt dimensions and layout to allow final detailed design work for foundations and any supporting structural steel and structures.
- All drawings and engineering data certified, sealed, and approved for construction.
- All required steel mill manufacturing reports and analyses, including material content and properties, as well as witness test reports and certificates of origin and completion.

The expediter's responsibility is to ensure that all of the information described above is completed in an acceptable manner and arrives from the vendor, supplier, or subcontractor in accordance with the dates and milestone activities specified in the purchase order or subcontract. One of the main reasons for delays in construction work is that vendors, suppliers, or subcontractors do not adequately compile and complete the necessary paperwork. Careful expediting of this situation will optimize the process.

Expediting Checklist

The following is a checklist of activities that are often performed by the expediting group when reviewing the status of equipment or materials in the fabrication cycle. On smaller projects, this activity could be performed by the project manager or the planner/scheduler.

- Review and understand the purchase order or subcontract requirements.
- Verify if vendor, supplier, or subcontractor has received purchase order or subcontract document.
- Obtain vendor's, supplier's or subcontractor's shop contract number or reference number for the article being purchased.
- Determine places of fabrication and sub-vendor's or supplier's name and address.

- Confirm that vendor, supplier, or subcontractor understands specifications and drawings.
- Obtain names of person(s) to contact; their titles, phone numbers, fax numbers.
- Inform vendor, supplier, or subcontractor if and when inspection activity is required.
- Determine whether detailed engineering and design data is required by the vendor, supplier, or subcontractor. If necessary, transmit engineering production schedule and any relevant additional engineering information to the appropriate organization(s).
- Check if the drawings prepared by the vendor, supplier, or subcontractor need approval by the state agency, owner, architect, engineer, or contractor.
- Establish location and person to whom drawings will be forwarded for approval or distribution.
- Determine whether welding procedures and fabrication method statements are required.
- Ensure that purchase orders have been released for fabrication to sub-vendors and third-party suppliers. Obtain unpriced copies if necessary.
- Obtain schedules indicating engineering, procurement, production, and fabrication sequences and major milestone dates.
- Review shop drawing and submittal schedule; check status on a regular basis.
- Obtain current promised delivery date(s); monitor and audit status on an ongoing basis.
- Confirm that the bill of materials or material takeoff (MTO) has been formulated and issued.
- Determine that required materials and equipment are available and in stock.
- Establish a list and audit those key purchase orders that need special expediting assistance and attention.
- Confirm that long lead placement of critical materials and equipment has been made and determine current status and promised delivery dates.
- Review approval methods related to fabrication shop drawings.
- Review turnaround time for approval of fabrication shop drawings and other related submittals.
- Determine status of production and shipping schedule of the main unit and all miscellaneous related equipment that is part of the complete purchase order.
- Review all necessary testing, inspection, and witnessing requirements.
- Confirm that shift work, overtime, and weekend work can assist in maintaining promised delivery date if necessary.
- Remain aware of cost ramifications of shift work, overtime, and weekend work.
- Determine whether material or equipment substitutions may be made to maintain or improve delivery schedule.
- Confirm that transportation and shipping instructions have been prepared and that delivery instructions are known by appropriate

project team personnel.

- Ensure that packing lists, operating manuals, start-up instructions, and various operating and maintenance manuals, together with necessary start-up spare parts and tools, are delivered to the construction site with equipment.
- Determine if vendor start-up personnel are required for start-up and equipment acceptance activities.
- Ensure that senior management is informed of any delays as soon as possible.

Expediting Status Reports

The forms shown in Figures 4.4a and 4.4b are typical expediting status reports. These reports may be modified and adjusted for other expediting applications.

Transportation

The responsibility of the transportation group is to convey materials and equipment to the project construction site. On smaller projects, this activity could be performed by the site manager or the planner/scheduler. The main factors to be considered in this important activity are listed below:

- **Safety.** Materials and equipment must be transported in a safe manner; the methods utilized should consider personnel safety and protection of the specific items being moved.
- **Price.** Transportation costs for a typical construction project can vary anywhere from 3% to 10% of the total project bottom line cost. Research should be undertaken so that shipments of materials and equipment are routed at the lowest acceptable price.
- **Timing.** The transportation group should be aware of the project schedule; materials should be shipped to the construction project prior to their scheduled delivery dates, or in the sequences that they are required.
- **Documentation.** The transportation group is responsible for the administration, collection, and subsequent delivery and maintenance of all documentation and paperwork associated with the transportation activities.

Planning for Transportation

Orchestrating transportation movements from the place of fabrication to the construction project location requires detailed front-end scheduling and planning. The project's plan should detail all the material and equipment items that require transportation activities. It is fundamental to the overall success of the construction project that the transportation plan consider the field construction need dates; the *need dates* reflect when materials and equipment are required at the construction site for their subsequent installation. If these need dates change or slip, the transportation group must be informed, since the overall project plans are predicated on receiving the materials and equipment on these predetermined dates.

Project location, project schedule, owner requirements, project funding, and potential unique transportation requirements are all elements

Initial Contact Report

ABC Constructors, Inc.

Project No. C-1234
Project Name Dobanol Chemical Fac.
Project Location Chicago, IL
Date 12/15/9X
Compiled by: Sharon Smith
Page 1 of 2

Fabricator's Name G+H Piping Inc.
Address 789 A Street, Anytown, PA 19010
Purchase Order No. 009 Revision No. 01
Revision Date 12/3/9X Promised Delivery Date 2/12/9X

Has vendor or supplier received copy of purchase order?
☑ Yes ☐ No ☐ Not applicable
If no, why not?

What is fabricator's or manufacturer's corresponding shop order number? S.N. 731-95
Has a detailed engineering schedule been prepared?
☐ Yes ☑ No ☐ Not applicable
If no, why not? 80% complete will be sent on 12/20/9X.

Has a detailed manufacture, fabrication, or projection schedule been prepared?
☐ Yes ☑ No ☐ Not applicable
If no, why not? Preliminary schedule has been prepared. Will be completed when detailed engineering schedule is received.

Does quoted or promised delivery date remain in effect?
☐ Yes ☑ No ☐ Not applicable
If no, why not? 1 week slippage due to delay of detailed engineering schedule

What is new date if quoted delivery date has changed? 2/19/9X
Does vendor or supplier understand all the submittal data requirements?
☑ Yes ☐ No ☐ Not applicable
If no, why not?

Location of the manufacturing/fabrication facility G+H Piping Inc., 789 A Street, Anytown, PA 19010
Contact Name and Phone No: John Casson, Operations Manager, 555-3721
What are the addresses and contact names of any sub-vendors or sub-suppliers?
N/A - No sub-vendors or sub-suppliers being utilized.

Figure 4.4a

Initial Contact Report (cont'd.)

Project No. C-1234
Project Name Dobanol Chemical Fac.
Project Location Chicago, IL
Date 12/15/9X
Compiled by: Sharon Smith
Page 2 of 2

Does the vendor or supplier know the delivery address of the finished product?

☑ Yes ☐ No ☐ Not applicable

If no, why not?

Does the vendor or supplier understand the invoicing procedure?

☑ Yes ☐ No ☐ Not applicable

If no, why not? Information contained with issued purchase order sent 11/23/9X

Contact names, telephone numbers, and FAX numbers of key personnel in the manufacturing/fabrication facility:

Position	Name	Tel. No.	FAX No.
Operations Manager	John Casson	555-3721	555-3819
Engineering Manager	John Price	555-3722	555-3819
Production Supervisor	Paul Boulton	555-3723	555-3819
QA/QC Manager	Beth Phillips	555-3724	555-3819
Purchasing Manager	Fred Wilson	555-3725	555-3819
Transport Manager	Ray Curl	555-3726	555-3819

Is manufacturer/fabricator issuing biweekly or monthly status reports?

☑ Yes ☐ No ☐ Not applicable

If no, why not?

Comments: Slippage of 1 week has occurred due to late delivery of detailed engineering schedule. New delivery date of 2/19/9X. Mr. Casson, Operations Mgr. has assured us that original delivery date 2/12/9X will be achieved. G+H Piping will work O/T at no cost to ABC Constructors to pull back 1 week slippage.

Distribution: John Casson, G+H Piping
Rick Rodgers, ABC Constructors, Site Mgr.
Sharon Smith
Project File: C-1234

Action:

Figure 4.4a (cont.)

Expediting Record/Report

Project No. ______________________
Project Name ______________________
Project Location ______________________
Date ______________________
Page _____ of ______________________
Prepared By: ______________________
Report No. ______________________

P.O./S.C. Delivery Date ______________________
Revised Delivery Date ______________________
P.O./S.C. No. ______________________
Equipment No. ______________________
Vendor/Supplier's Name ______________________
Address ______________________
Telephone ______________________
FAX No. ______________________
Contact Name ______________________
Title ______________________

Expediter's Name ______________________
Contact Date ______________________
Contact Time ______________________

Ref.	Comments	Action By

Comments: ______________________

Distribution: ______________________

Action: ______________________

Figure 4.4b

that must be considered in the transportation plan. Specific factors that may influence planning considerations for transportation have been discussed in Chapter 2.

Large or Special Loads of Equipment and Materials

Some of the types of large or special loads that the transportation group might be responsible for include:

- Large pieces of process equipment (distillation towers, vessels, compressors, conveyors, storage tanks)
- Large elements of construction equipment, such as cranes, excavators, and tunneling equipment
- Fabricated modules or pre-assemblies
- Fabricated structural steel components and sections
- Large precast concrete beams, columns, or panels
- Delicate and/or sophisticated electronic control equipment

Large, difficult, or special loads of material or equipment necessitate efficient and complete communication with many diverse groups. In such cases, the transportation group must coordinate and communicate with the following groups:

- The engineering or design group performing the detailed engineering, to determine whether the equipment item can be broken down into small or mid-sized movable elements
- The manufacturer/fabricator or supplier, for the same reason
- The transport or freight company, to determine the number of loads to be moved
- Any regulatory authorities, city or state highway departments that are responsible for roads, highways, etc., on which these articles will be transported
- The project management group at the project construction site

Regulatory Issues: Certain restrictions, rules, and practices are enforced by agencies such as city highway departments, state transport agencies, and foreign government agencies. Each locality has different rules and regulations regarding the transportation and movement of large or special loads of materials and equipment. Certain permits, approvals, and documentation may be required for the transportation of large materials and equipment through a particular city, state, or country. Therefore, the transportation group must engage in careful planning and coordination prior to the movement of any items to or from a particular area.

Freight Forwarders

Freight forwarders are generally used by organizations that have complicated transportation requirements. A freight forwarding organization selects the most practical and cost-effective method available for moving materials or equipment to a specific domestic or international location, and prepares all of the necessary transportation documentation. Many times a freight forwarder will combine smaller shipments to optimize transportation costs. Freight forwarders deal on a day-to-day basis with rail, road, air, and ocean transportation companies; they can be considered experts in the transportation field.

The services provided by freight forwarding organizations can vary in many ways, depending on the requirements of the organization needing materials and equipment at a specific location. Listed below are some of the tasks that freight forwarding companies perform on a routine basis:

- Arrange for booking and blocking out of space with freight/transportation companies (sea, highway, air, and rail).
- Collect and distribute all documentation: packing lists, manifests, shipping papers, and so on.
- Arrange for the physical delivery and receipt of supplies and equipment at the final port or destination.
- Prepare export documentation, including any required customs clearance certification and permits.
- Check for shortages of and damage to materials and equipment prior to and after unloading process.
- Prepare any documentation for shipment of supplies from original place of manufacture to final destination.
- Control, monitor, and audit all shipments and deliveries to project construction site from a port of entry.
- Control, monitor, and audit all freight forwarding activities: items related to insurance, damage reports, permits, customs inspection, customs fees, duties, temporary storage, bonded warehousing requirements, and all necessary import documentation.

Planning for Foreign Transportation

Transportation of materials and equipment to foreign construction sites requires additional logistical considerations and support. Late delivery or damage to equipment while in transit can create major problems, in terms of both the project's budget and completion date. To ensure that materials and equipment arrive at the foreign location on schedule and undamaged, the material and equipment need dates must be established early in the planning phase of the construction project. Listed below are factors that might influence the transportation of materials and equipment to overseas construction projects.

- Adequate ocean or air transportation is not always available on a regular basis. Available space on vessels and aircraft is often booked by other organizations, sometimes many months in advance. Careful planning should be made for shipment of large or complex materials or equipment.
- U.S. customs and export requirements, together with all the necessary documentation and data, will need to be fully understood and complied with.
- Stringent requirements for packing and crating of materials and equipment may be required.
- Every overseas country has unique customs requirements: duties, tariffs, taxes, permits, and licensing requirements.
- Certain ports have limitations concerning off-loading, storage capacity, docking capacity, ship's draft and keel clearance, heavy lifting equipment, lay-down areas, and warehouses.
- Local road and rail transportation may not be adequate. Bridge clearances, tunnel sizes, or bridge weight limitations may need to be considered.
- Off-loading equipment and adequate storage facilities must be

in place at the construction site prior to the arrival of materials and equipment.

- The majority of all overseas countries use the S.I. metric system of measurement; transportation forms and paperwork must be designed for meters, kilograms, and so on.

Each country has its own requirements and regulations regarding transportation of materials and equipment, so the above list is in no way exhaustive. Permits and monetary considerations such as taxes and duties are among the most crucial factors when foreign transportation is being considered.

Export Permits: The U.S. Department of Commerce controls, monitors, regulates, and audits the export of all materials and equipment. A general export permit or license is usually required for normal construction materials and equipment. Special permits and licenses are required for high technology equipment or materials. The Department of Commerce maintains a detailed listing of these high technology items; they include certain advanced weapons systems and computer equipment and software. Rarely would normal construction materials and equipment fall into this high technology category.

All details related to the export of materials and equipment should be compiled and forwarded to the U.S. Department of Commerce as early as possible in the project execution cycle to ensure against any possible delays in transportation and delivery.

Import Permits: The U.S. Department of Commerce has an equivalent agency in every overseas country. These government agencies are charged with at least as many responsibilities as the U.S. Department of Commerce. The import of materials and equipment for a foreign construction project is almost always subject to government regulations, reviews, surveillance, and audits. The transportation group must determine the best methods to import needed materials and equipment.

Contact should be made with the foreign country's U.S. embassy. These foreign embassies can usually give direct assistance or, if necessary, provide the name of the correct government agency or department in the country to contact.

Tariffs, Duties, and Taxes: In many cases, foreign countries impose tariffs, duties, and taxes on the importing of items such as construction materials and equipment. These costs can range from 0.5% to 35% (or more, in certain circumstances) of the purchase price of the materials and equipment. Such tariffs, duties, and taxes are often used to protect the country's domestic manufacturing industry and to discourage the purchase and importation of materials and equipment from foreign manufacturers. The transportation group should determine what tariffs, duties, and taxes will be required in a particular country, when and to what government agency they are to be paid, and what method of payment is required.

The transportation group should maintain an ongoing file of all pertinent foreign tariffs, duties, and taxes. An excellent resource for this particular subject is the book entitled *Exporters Encyclopedia* which is published annually by Dunn & Bradstreet International.

Transportation Checklist

The following is a checklist of transportation considerations related to domestic and overseas construction projects. This checklist covers the drafting of transport contracts and general transportation activities.

- Use standardized forms and documents whenever possible.
- Ensure that the transportation company is aware of all transportation obligations and requirements, including specific delivery dates.
- Ensure that the bill of lading or airway bill that will govern the movement of materials or equipment to the construction job site is completed.
- Ensure that payment terms are fully understood by all parties.
- Taxes, tariffs, import duties, and all other related fees must be detailed, along with an indication of the party responsible for payments.
- Ensure that contracts and purchase orders include the following:
  a. pricing details, including any discounts or special terms
  b. a detailed list of all materials and equipment to be transported, including pick-up, loading, or commencement location, final destination, and delivery dates
  c. a force majeure clause for wars, strikes, floods, earthquakes, acts of God, etc.
  d. a standard indemnification clause to protect the owner and contractor in the event that the transportation company cannot meet its obligations
  e. a statement indicating which party is responsible for obtaining insurance coverage and what level of coverage is needed
  f. provisions for future modifications to the contract
  g. a statement that the contract may not be assigned to another party without the written consent of the owner or contractor
  h. a statement indicating which state's or country's laws will prevail in a dispute
- Establish a transportation procedure by which all transportation movements are governed.
- Prepare all the required freight routing directions for the transportation of materials and equipment to their destinations.
- Prepare a list of materials and equipment that require special packing; some delicate items may require careful, expensive packing work to protect them during transit and delivery.
- Develop scopes of work, compile a list of acceptable bidders, and issue bid documents to acceptable transportation companies.
- Assist in the bid selection process; monitor and audit the performance of transportation companies, freight forwarding agents, air freight companies, rail transportation companies, courier services, and warehousing companies.
- Coordinate and keep the project team informed about significant events related to transportation activities; issue bi-weekly or monthly transportation status reports.

The Transportation Status Report

Figure 4.5 is a transportation record that may be used or modified for future transportation record applications.

Determining the Success of a Construction Project

Research and experience have indicated that successful construction projects have largely depended on detailed front-end planning, scheduling, and coordinating related to quality, inspection, expediting, and transportation considerations. There are many elements related to these activities that must be considered during the course of a project. Many of these elements have been discussed throughout this chapter, but they deserve reemphasis.

Knowledge is an important tool. All project team members must be informed of the owner's requirements related to quality, delivery, and cost. In addition, vendors, suppliers, and subcontractors must be given clear and concise scopes of work. Materials and equipment that will require inspection and testing must be recognized and discussed by the appropriate parties before the project commences.

The construction purchasing management group must be able to communicate to suppliers, vendors, and subcontractors that avoidance or correction of problems early in the construction process is beneficial to all parties. Early planning in this regard can minimize delays and reduce the possibility of expensive rework.

To perform his or her job effectively, the expediter must continually research the status of the fabrication or manufacturing process. No question is ever insignificant; only by asking many questions can the expediter ensure that the required completion date will be achieved. The expediter should never accept any slippage in the schedule. Overtime, shift work, weekend work, possible material substitutions, and acceleration premiums to vendors or suppliers may all be considered to uphold promised delivery dates.

Transportation and delivery requirements should also be carefully planned early in the project. Producing equipment and materials at the correct time may be a difficult task but the safe delivery of these items is crucial to the success of any construction project. The transportation plan should be based on project milestones and material and equipment need dates. Knowledgeable transportation staff will always have a positive impact on the construction project.

Finally, the success of any construction project is heavily influenced by efficient and accurate reporting. Status reports are vital information sources to the project management team. Proper reporting provides the knowledge that is necessary to address problems and keep the construction project moving ahead smoothly.

Transport Record

ABC Constructors, Inc.

Project No. C-1234
Project Name PTM Facility
Project Location Nice, France
Date 5/21/9X
Page 1 of 1
Prepared by: Sharon Smith
Report No. 2

Purchase Order/Contract No. P.O. 017
Supplier/Vendor Wilson Compressor & Fan Corp. Record No. 1 of 1
Contact Debbi Herr, Asst. Purchasing Mgr.
Telephone No. 555-9329
Address 123 Main Street, Anytown, NY 10500

Amount or Value of P.O./Contract $ 23,100.00
Transport Movement/Shipping Record T-1071

Shipment/ Ref. No.	Material/ Equipment	Gross Weight	Transport/Air Ocean Company	Point of Origin Port	Bill of Lading No.	Freight Cost
001	Compressor C-001	2150 lb.	XYZ Shipping Ltd.	Boston, MA	B.L.-19113	$ 1953
002	Compressor Housing & Frame	750 lb.	↓	↓	B.L.-19114	$ 873
003	Compressor Lube-Oil System	350 lb.	↓	↓	B.L. 19115	$ 485
004	Compressor spare parts	255 lb.	↓	↓	B.L. 19116	$ 390

Comments: XYZ Shipping Ltd. vessel "Ocean Ranger" departs Boston, MA on 5/24/9X. Arrives Port of Nice, France 6/2/9X. XYZ Shipping Ltd. to contact Mr. Bill Jones, Site Manager, ABC Constructors, Inc. at 011-33-93-31-07-39, when equipment has passed through customs clearance.

Distribution: Bill Jones, Sharon Smith, Debbi Herr, XYZ Shipping, Project File

Action:

Figure 4.5

Chapter 5

Contracting, Subcontracting, and Administration

Chapter 5

Contracting, Subcontracting, and Administration

A construction project is a complex undertaking that requires detailed coordination among various individuals, parties, and organizations. Large sums of money are required for the procurement of construction materials, equipment, and labor; indirect costs include the purchase of land, financing, and engineering and design services. The process of coordinating, managing, and administrating the financial and installation aspects of a construction project requires considerable management experience and business skills.

The owner of a construction project cannot rely on a handshake or verbal agreement in today's business environment. Binding contracts or agreements are the norm in the construction industry as they are in all other industries. Construction professionals should ensure that all project contracts contain the following elements: detailed scope of work; quality of materials and equipment to be used; time frame for performance and completion of work; and any qualifications, exclusions, or alternatives to the contract. A comprehensive contract can serve as general insurance against major problems on the project, or at least as protection for the parties involved should problems occur.

Unique Characteristics of the Construction Industry

The construction industry is different in many ways from industries such as manufacturing, chemical, oil, steel, or transportation. Every construction project is one of a kind, never repeated under the same conditions or time frame. Because of this uniqueness, each project requires detailed planning, beginning very early in the development phase. The average construction project is completed in 12 to 24 months, as opposed to a manufacturing process that might have a life cycle of up to 10 years.

Listed below are some of the key elements that make construction work unique.

- Construction projects are planned and executed under very strict time constraints.
- The design and engineering effort typically cannot be utilized again for other projects.

- Materials and equipment are usually subject to a great deal of the waste, theft, and damage because of the temporary nature of construction site or storage facilities.
- Labor is usually fragmented into various work crews or subcontractors that do not always work in a coordinated manner.
- Problems can arise on the construction project when work is performed by both union and nonunion labor. Strikes, walkouts, and, in extreme cases, sabotage to completed work can occur.
- As construction workers complete a job, they are working themselves out of a job (their job is over when the work is done). In certain situations, this can cause low productivity and high absenteeism.
- More than any other industry, construction work is subject to weather conditions. Bad weather can have a serious effect on the completion date of a construction project.

All of these factors should be considered and fully understood by the owner before a construction contract is developed.

Participants in the Construction Project

All construction projects – warehouses, schools, hospitals, manufacturing facilities, chemical plants, shopping malls – require the special knowledge and skills of certain key participants to make the project concept a reality. The key participants are usually referred to as the *project team*, and consist of the owner, the designer (engineer or architect), and the general contractor.

The Owner

The owner(s) may be a private individual or a partnership, a corporation, or a government agency. The owner should have a clear title to the property on which the project will be built, and should be able to make timely decisions regarding the construction project.

Once the decision to construct a building has been made, the owner, sometimes with the help of a design firm, conducts basic feasibility and engineering studies to develop overall project parameters and guidelines. The basic considerations in the owner's planning process are:

- Type of facility or building to be constructed
- Size and function of the facility or building
- Date when the facility or building must be completed and in operation
- Life cycle of the facility or building
- Quality and reliability of the facility or building
- Detailed scope of work to be performed
- Conceptual budget or estimate of the proposed new facility or building
- Start and finish dates of major milestones related to the execution of the construction project

The owner usually enters into a contract directly with the designer of the project. In some cases, the owner might also hire a construction manager to monitor the design and construction process. On the vast majority of construction projects, the owner is involved in two contracts: with the designer and with the general contractor.

Figure 5.1 depicts the owner's typical contracting process.

The Designer

The designer is usually an architect or engineer who must be licensed by the state in which the construction project will be built. The designer produces the conceptual and detailed design documents, project specifications, and completed drawings that will be used throughout construction of the building or facility.

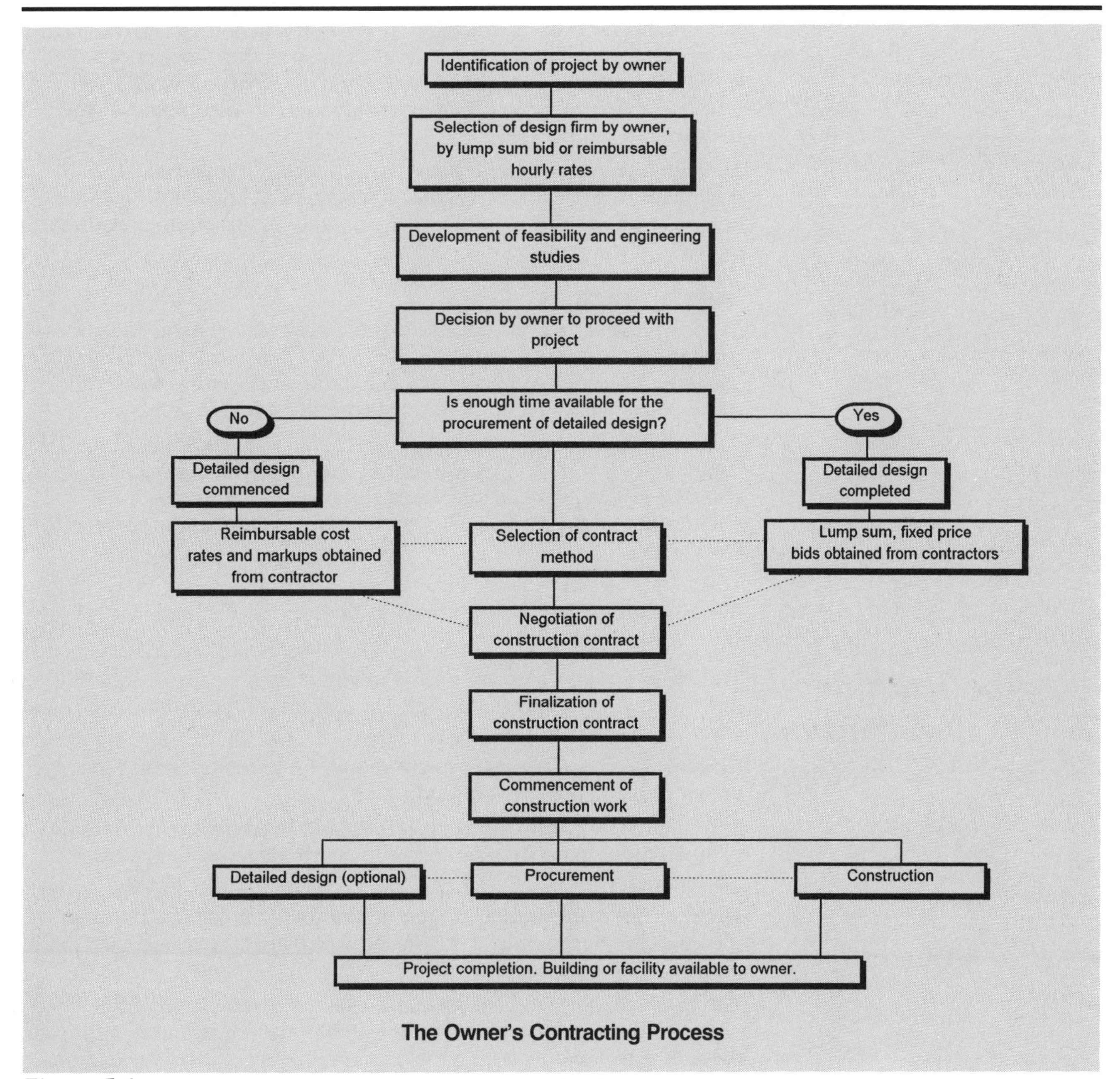

Figure 5.1

Upon completion of the design, the designer may assist the owner in advertising for and selecting the contractor to be used for the construction project. During construction, the designer usually makes periodic site visits to ensure that the materials and work are in compliance with the specifications and drawings. These periodic visits should not be interpreted as full-time inspection or quality control surveillance.

The Contractor

The constructor is usually known as the general contractor. The general contractor brings together all elements of the construction process: the materials, equipment, and construction labor and services. The general contractor can be a private individual, a partnership, a joint venture, or a corporation. In certain states the general contractor must be licensed to be able to perform construction work. The most successful general contractors typically possess the experience, resources and the ability to complete the construction work in an efficient, safe, timely, and cost-effective manner, in compliance with the owner's bid package.

After the owner and the designer have selected a qualified general contractor, the owner enters into a contract directly with the general contractor. The contractor then performs the construction work that was detailed in the owner's bid package.

Subcontractors

In today's construction industry it is the norm for general contractors to hire specialist subcontractors for certain key work elements of the construction project. Electrical, mechanical, and roofing work are some of the work elements that are typically subcontracted.

Subcontracts are established between the general contractor and the subcontractor without involvement by the owner. The owner's contract with the general contractor makes the general contractor totally responsible for the performance and the completion of the construction work.

The traditional contractual arrangement, including the team members discussed in the previous sections, is depicted in Figure 5.2.

The Traditional Construction Method

The project team's primary goal is to construct a facility or building that meets the needs of the owner. These goals center around three main concerns:

Budget: The construction project should be executed within the pre-established project cost budget.

Schedule: The construction project should be completed within the planned time frame determined in the early stages of the project planning cycle.

Quality: The construction project should be completed according to the quality requirements of the owner, which are established early in the planning phase of the project.

The project team needs to plan and execute certain key actions to meet the three primary goals. In the traditional construction approach these actions are:

- Identification of owner requirements
- Selection of design consultant; preparation of preliminary and detailed design
- Assessment of space, facility configurations, headcount needs, and so on
- Feasibility studies, evaluations of options (new facility or renovation of existing facility), and cost estimates for funding requirements
- Selection of the most attractive building or facility configuration (i.e., size or layout)
- Preparation of construction bid package; receipt of competitive lump sum bids or cost reimbursable bids from qualified contractors
- Selection of contractor; commencement of construction

Elements of the Construction Contract

A *contract* is a legally enforceable agreement between two or more individuals, parties, or organizations. The agreement defines the obligations and rights of the parties involved.

A construction contract is an agreement between the owner and the contractor in which the contractor agrees to construct a facility or building in accordance with certain contract documents. These contract documents consist of drawings and specifications that are usually prepared by an independent third party such as an architect or engineer. The contractor will agree to construct the facility or building in a specific time frame for a stipulated price. If the owner and contractor agree to this arrangement, the end result is a construction contract.

Contracts can be oral, made by spoken agreement, or they can be written, in which both parties sign to indicate their acceptance. Oral

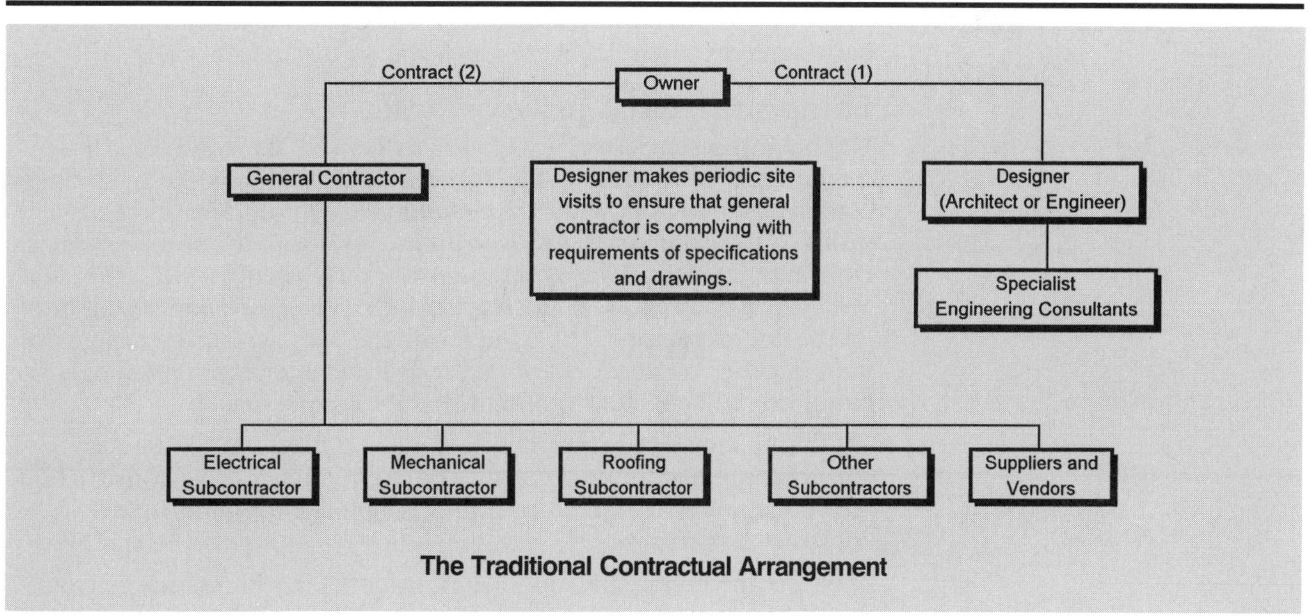

The Traditional Contractual Arrangement

Figure 5.2

agreements are rarely used for construction projects because they are difficult to enforce in the event of a dispute. Owners of construction projects should have in place a valid written contract that protects their financial, legal, and business interests.

Construction contracts can be formulated in many ways depending on the requirements of the owner and the contractor. The most successful construction contracts have four basic qualities. They should be *clear*, *concise*, *complete*, and *flexible*. Flexibility is important because owners often request changes during construction.

Along with those four qualities, other elements must be in place to make a contract valid and legally enforceable. The most essential of these elements are listed below.

- **Agreement** is the mutual consent to the terms and conditions stipulated in the contract. This is usually referred to as "offer and acceptance" – an offer is made by one party and agreement to the terms implies an acceptance of the offer.
- **Competent Parties** refers to the concept that each party must be legally capable of executing his or her obligations in the proposed contractual arrangement. A contract cannot bind an individual who does not have the legal capacity to perform such an obligation.
- **Proper Subject Matter** refers to the purpose of the contract. The end product of the contract must have a lawful purpose and a proper subject matter that conforms with the law. A contract is illegal if the subject it addresses is prohibited by statute or law.
- **Valid Consideration** involves compensation or payment from one party to another for work performed.

Other elements that should be part of a valid contract are contract dates, correct monetary values, and correct signatures of principals and witnesses along with any required seals or attestations.

Types of Construction Contracts

The following is a summary of the principles, advantages, and disadvantages of three basic types of construction contracts. In addition to these, there are five contract types that are variations to the three basic agreements. All of these types will be discussed.

Lump sum, fixed price contracts

When a lump sum, fixed price contract is used, the contractor has total budget responsibility for the construction project; he or she must complete the project within the stated price. An underrun means a profit to the contractor; an overrun means a loss to the contractor. Drawings, specifications, and design work are usually 100% complete. The owner contracts separately with the architect or engineering firm for the design package. The owner, with the assistance of the design firm, then qualifies contractors and requests lump sum, fixed price bids based on the architect's or engineer's design package.

The lump sum, fixed price contract may be used for commercial construction, institutional construction, civil engineering construction, residential construction, and selective manufacturing/industrial construction projects.

To select the contractor, the owner, sometimes with assistance from the architect or engineer, prequalifies possible contractor candidates.

Selected contractors are requested to submit a lump sum, fixed price bid based on the owner's design package on or before a fixed bid date. In most cases, the construction project is awarded to the lowest bidder who meets all the requirements of the instructions to bidders and the bid documents.

In lump sum, fixed price contracts, the competitive nature of the marketplace is a cost advantage to the owner. The owner knows his or her total projected cost expenditure at the commencement of construction, and reduces his or her risk exposure by transferring the risk to the contractor. The owner, however, minimizes his or her overall involvement in the day to day decision making of the project. Construction productivity and cost containment are maximized because the contractor works within a fixed, pre-established budget.

There are certain disadvantages to lump sum, fixed price contracts. Construction work cannot usually commence until the design effort is complete. There is an overlap of certain project responsibilities of the designer and the contractor. Changes requested by the owner during the construction effort can result in disputes; cost increases and delays to the project's end date can develop into an adversarial relationship between the owner and contractor. Therefore, unit prices for extra or deleted work should be included in the bid package; the contractor should be requested to provide this information with his or her bid submission.

Cost reimbursable contract, or cost plus fee contract

A cost reimbursable contract gives the owner cost and budget control responsibility for the project.

Drawings, specifications, and design work are usually incomplete in this case. Contractors are often unwilling to submit lump sum, fixed price bids on work that is not fully defined because it is almost impossible to accurately estimate the cost of undefined work. However, the owner is often unwilling to wait for the design to be completed if he or she has to have the building or facility in operation as soon as possible. The owner is sometimes performing emergency repair work to his or her building or facility; typically the scope of the project is not completely defined.

Cost reimbursable contracts are usually used for large manufacturing/industrial construction projects, in which the detailed design has not been fully completed, or when emergency repair work has to be completed. To select the contractor, the owner usually negotiates with qualified contractors on billing rates, overhead percentages, material markups, and profit fees. The owner selects the contractor who seems best qualified and who apears to be the most cost-effective to perform the construction work.

When cost reimbursable contracts are used, there is a less adversarial relationship between the owner and contractor. Construction can commence before the design effort is complete, and changes requested by the owner can be easily accommodated by the contractor. However, many times the owner must provide support personnel to review and audit engineering, procurement, and construction activities. Cost reimbursable contracts are typically more expensive than lump sum, fixed price contracts. Another disadvantage is that the owner doesn't know his or her final cost until the project is completed.

There are several adaptations of cost reimbursable contracts:

- In *cost plus percentage fee* contracts, all costs are reimbursable, and the contractor receives a stipulated profit percentage fee.
- *Cost plus fixed fee* contracts are very similar to cost plus percentage fee contracts, but the profit is a fixed amount.
- In *time and material or daywork rates* contracts, hourly billing rates that include overhead and profit are paid to the contractor by the owner. Materials are paid at cost, plus a percentage to cover overhead and profit.
- In *cost reimbursable converted to lump sum fixed price* contracts, at some point during construction the owner asks the contractor to submit a lump sum bid for completion of the work.
- In *direct labor cost plus fixed percentage for overhead* contracts, payroll records are used as records of costs that the owner will reimburse.

Negotiated Contracts

The operating principles, selection process, advantages, and disadvantages for negotiated contracts vary considerably from one project to another. The owner, with assistance from the architect or engineer, selects a qualified contractor to perform the construction work. The owner and the contractor then negotiate a compromise contract that is satisfactory to both of them.

A negotiated contract can be a lump sum, fixed price contract, a cost reimbursable contract, or a combination of both. Negotiated contracts may be used for all categories of construction, but are not commonly used for public works or institutional construction projects.

Although historically negotiated contracts have been used mainly for manufacturing and industrial construction, in recent years they have become more widely used by private industry owners. In the future, negotiated contracts will be used more extensively for residential, commercial, and civil engineering construction work.

Unit Price Contracts

The unit price construction contract is based on estimated quantities of well-defined items: cubic yards of excavation or concrete, tons of structural steel, and so on. The owner qualifies contractors and requests unit prices. The contractor submits a unit cost value to perform each item of work. The lump sum unit prices include the contractor's direct and indirect costs – all labor, materials, and construction equipment, together with overhead and profit. The owner usually selects the contractor who has submitted the most attractive unit prices.

The quantities are often difficult for the contractor to estimate accurately prior to the contract award, because drawings, specifications, and design work are typically incomplete. Therefore, in most cases the contract provides protection to the contractor if actual quantities vary significantly from those shown in the bid package. The contractor is reimbursed for the final actual quantities installed.

Unit price contracts are widely used on civil engineering construction work. When a unit price contract is used, construction work can commence before the exact quantities are known. Open competition among contractors can be maintained, and reimbursement and payment routines are well defined.

Unit price contracts do have some drawbacks. The owner usually requires a permanent field staff to audit the contractor's monthly and final invoice submissions. Also, contractors usually inflate unit prices of low quantity work items, hoping that the item will increase in quantity when the work is fully defined. This situation can seriously jeopardize the owner's construction budget if he or she does not carefully review the unit price values prior to contract award.

Guaranteed Maximum Price Contracts

Many owners are not pleased with a cost reimbursable contractual arrangement because the final construction expenditure is not known until the completion of the project. A guaranteed maximum price, or GMP, is a form of contract in which the contractor guarantees that the project will be constructed in accordance with the scope of work as outlined by the owner and that its cost will not exceed a predetermined target price. The contractor determines the target price based on specifications and drawings that are sometimes less than 15% complete and adds a fixed fee to the target price. If the cost of the construction work exceeds the target price, the contractor pays for the overage.

In a GMP arrangement, the owner is assured that the construction project has a budget cap that will not be exceeded. Many GMP contracts contain a bonus or incentive clause that can assist in keeping expenditures below the budget cap. This clause usually states that the contractor will share in any savings below the guaranteed maximum price or budget cap. It typically allows the contractor a fixed percentage of the savings realized on the construction project. This percentage savings can usually vary from 5 – 50%.

One advantage of GMP contracts is that they are established without the time-consuming preparation of detailed specifications and drawings. This is an excellent arrangement for an owner because schedule and cost goals can be realized. The owner is assured that the contractor is motivated to obtain maximum efficiency in the field and to make or recommend relevant savings to the GMP where possible. In addition, the contractor has a major incentive to improve his or her productivity and performance since he or she usually shares in the project's cost savings. Finally, the owner retains options to approve and be involved in all major project decisions.

When using a GMP contract, the contractor usually protects himself with a conservative first-pass cost, believing that the owner will negotiate his or her price down on the first price submission. Therefore, a drawback to GMP contracts is that the contractor's fee and contingency fund is usually higher than on competitively bid fixed price, lump sum contracts.

Bonus/Penalty Incentive Contracts

A bonus/penalty incentive contract allows the owner to offer a monetary bonus to the contractor for early completion of the construction project. For example, the owner may benefit greatly if the facility or building is completed earlier than scheduled. To this end he or she may offer the contractor a bonus for each day or week saved in the schedule. However, just as a bonus can be offered, a penalty for late delivery can be deducted from the contractor's lump sum fixed price bid.

Management Contracts

In a management contract, a contractor is hired to manage the construction work on behalf of the owner. This method is typically used by owners who have limited in-house engineering resources. The owner selects the contractor because of his or her construction and administration abilities.

For a fixed or percentage-based fee, the selected management contractor will coordinate and manage the construction effort. The management contractor usually will not perform any of the construction work, but will be responsible for construction management, estimating, budget control, planning, placing of subcontracts, purchasing of materials and equipment, safety matters, and coordinating and monitoring the entire construction effort.

Design/Build Contracts

A design/build contract provides all of the services (design, procurement, and construction) to complete a building or facility for an owner. Design/build contracts are sometimes referred to as "turn-key" projects, because the owner, when the construction work has been completed, may turn the key of the main door of his or her new facility or building and take immediate possession of a fully completed building or facility that matches his or her requirements.

In design/build construction contracts the owner specifies his or her requirements (e.g., number of widgets to be manufactured per day, number of guests staying in the proposed hotel, or tons of cement to be produced per year in the new cement manufacturing plant). There are of course countless variations to these needs. The owner will also stipulate the date by which the facility or building must be constructed, the quality requirements for materials and equipment, and any other requirements he or she feels are necessary.

The contractor designs, procures, and constructs the facility within the specified parameters. The owner will usually have one contract with the contractor; the contractor may have a number of contracts and subcontracts with other parties.

A design/build contract may be a fixed price, lump sum contract, a cost reimbursable contract, or a negotiated contract.

Contract Selection

Combinations and variations of the basic contracts can be formulated depending on the needs of the contracting parties. Each type has distinct advantages and disadvantages.

The selection of an improper or inadequate contract can have significant effects on the completion and final cost of the construction project. In addition to the owner's primary needs, the contractor's goals should be considered. For example, for construction projects that run for more than two years, a contractor is taking a great financial risk if he submits a lump sum bid. Because high inflation rates can affect the cost of materials, equipment, and labor, lump sum contracting will always be prone to many risks.

The owner must carefully consider which contracting strategy to use. Construction projects with a short duration and a well-defined engineering and design scope should, in most cases, use a lump sum bid method. Longer construction projects with engineering and design work that are less than 75% complete should, in most cases, use a cost reimbursable contract or a negotiated contract.

In general, the factors that an owner should consider prior to selection of the contract type are:

- Owner involvement in the design and construction effort
- Time available to prepare scoping studies and feasibility studies
- Special considerations for extremely large, complex construction projects of variable scope and magnitude
- Duration and length of project execution
- Use of innovative design based on new technology, equipment, and materials
- Time available to study the documents, evaluate market conditions, and study the site location
- Time available to prepare a complete project document package
- Schedule limitations and completion date
- Remote, hostile, or rigorous site locations
- Considerations for a "first-of-a-kind" construction project
- Budgetary limitations on the construction project
- Ability to obtain competitive bids.

After reviewing these factors the owner should be in a position to determine which form of contract would best suit his or her specific construction needs.

Contractor Selection

The owner may use one of three basic approaches in selecting a contractor.

Closed Bidding

In the *closed bidding* approach, the owner, usually with the assistance of the architect or engineer, selects from three to perhaps as many as ten qualified contractors. The owner issues an invitation to bid to all of these contractors. To bid competitively on the work, the contractor must be able to comply with the following requirements:

- Be licensed by the city and/or state in which the work will be performed
- Be able to demonstrate that he or she has performed work similar to the proposed project
- Have a reputation for quality construction work

The closed bidding process is typically used by private industry owners. Closed bid construction projects are not normally publicly advertised. The contract is usually awarded to the lowest bidder who satisfies all the requirements of the owner's bid package.

Open Bidding

Open bidding is the standard operating procedure for public works projects. Public works construction includes federal, state, county, city, and town construction projects financed by public revenues and bond issues. As its name implies, open bidding is open to all qualified contractors who want to bid on the construction project. The owner, usually with the assistance of the architect or engineer, analyzes the bids, negotiating an agreement or contract with the lowest bidder.

Open bid construction projects are usually advertised in newspapers and/or trade magazines. The advertisement usually contains a brief description of the construction project as well as information regarding where bid documents can be obtained and reviewed, and where and when the bid is to be submitted.

Negotiated Contracts

On certain construction projects, the owner may forgo either the competitive closed bidding or the competitive open bidding selection process. This decision might be made because of the owner's schedule requirements. In this situation the owner, usually with the assistance of the architect or engineer, selects two to four contractors. These contractors should be able to satisfy the licensing, quality, and experience requirements listed in the description of closed bidding. Negotiated contracts are usually used by private industry owners. As stated earlier in this chapter, the negotiated contract can be either a lump sum fixed price contract or a cost reimbursable contract, or a combination of both.

Award of the Construction Contract

In many cases the award and placement of construction contracts and subcontracts are very similar to the award and placement of purchase orders. The following checklist may be modified for either an owner or a general contractor, as the situation dictates.

- Prepare notices and advertisements for bids
- Prepare bid package documents
- Review bid documents with legal department
- Establish bidders list
- Screen contractors, subcontractors
- Prequalify contractors, subcontractors
- Deliver bid documents and addenda to contractor and subcontractors
- Perform pre-bid conferences
- Furnish relevant information to all bidders if necessary
- Contractors, subcontractors prepare bids
- Receive bids
- Open bids
- Perform technical evaluation
- Perform commercial evaluation
- Prepare bid tabulation reports
- Recommend short-listed bidders
- Negotiate
- Select contractor, subcontractors
- Prepare contract documents
- Award contract
- Hold pre-construction or kickoff meeting
- Ensure permits, insurance, and performance bonds, etc., are in place prior to commencement of work.

The Construction Contract Document Package

For any type of contract, the correct preparation and review of the construction contract document package are essential ingredients of sound contracting practices. The absence or deficiency of elements related to the construction contract document package will heighten the risk of contractual, financial, and legal disputes and subsequent claims. For any construction contract or agreement to be valid, the obligations of each party must be completely spelled out. The legal rights and possible remedies in the event of a dispute must be fully described. A detailed scope of work should also be part of the contract documents.

The following is a list of the elements included in a typical construction contract document package. An owner would create this package with the assistance of the architect or engineer.

Bid requirements

- Invitation to bid
- Intention to bid confirmation (completed within 5 days by contractor indicating his or her intent to bid)
- Instruction to bidder
- Bid form
- Attachment to bid
- Scope of work statement
- Information available to bidders (soil bore logs, property surveys, site data, and so on)

Contract forms

- Form of contract or agreement
- Addenda
- Certificate of insurance requirements
- Bonding requirements, bid bonds, payment bond, performance bond

Specifications

This list is based on the MasterFormat classification, which is widely used by architects and engineers in all types of construction (except civil engineering and manufacturing/industrial construction).

- Division 1 General Requirements
- Division 2 Site Work
- Division 3 Concrete
- Division 4 Masonry
- Division 5 Metals
- Division 6 Wood and Plastics
- Division 7 Thermal and Moisture Protection
- Division 8 Doors and Windows
- Division 9 Finishes
- Division 10 Specialties
- Division 11 Equipment
- Division 12 Furnishings
- Division 13 Special Construction
- Division 14 Conveying Systems
- Division 15 Mechanical
- Division 16 Electrical

On institutional construction projects, architects and engineers sometimes use the Uniformat construction classification. Many government agencies use this construction classification.

- Division 1 Foundations
- Division 2 Substructures
- Division 3 Superstructures
- Division 4 Exterior Closure
- Division 5 Roofing

- Division 6 Interior Construction
- Division 7 Conveying Systems
- Division 8 Mechanical
- Division 9 Electrical
- Division 10 General Conditions and Profit
- Division 11 Special
- Division 12 Site Work

Contract drawings

- List of all project drawings

Instructions to Contractors

In closed, open, and negotiated bidding the owner, usually with the assistance of the architect or engineer, will issue to the bidding contractors the construction contract document package. Contained within this package will be a document entitled Instructions to Contractors, which describes the operating procedures of the bidding evaluation. The Instructions to Contractors should contain information regarding the following:

- Request for proposal cover letter
- Name and address of the owner
- Name and location of the construction project
- Names and addresses of the architect and engineer
- Bid form detailing the bid price
- Detailed description of the scope of work to be performed
- List of key milestone dates (e.g., commencement of construction, completion of construction)
- Place, date and time for delivery of bids
- Security procedures related to bids
- List of contract drawings
- List of applicable specifications
- Location and procedure for review of construction contract document package
- Prequalification questionnaire document
- Description of specific site conditions and site data
- Owner's requirements related to safety and work procedures

When the contract documentation package and the Instructions to Contractors are received by the contractors, they start producing their bids on the subject work.

The Instructions to Contractors should also contain a statement giving the owner the right to accept or reject any or all bids. This statement should also state that the owner further reserves the right to award the work to the contractor who will best serve the interests of the owner.

Bidding

Many factors determine the availability of bidding opportunities to contractors. Construction projects come to the construction marketplace in no set pattern or time cycle. Of course, not all contractors can bid on every construction project and not all construction projects are appropriate for every contractor. The senior management of any construction organization must determine which bidding opportunities the company will pursue. In making such a decision, he or she may consider many of the following factors:

- Type of construction
- Location of the construction project (distance from home office or branch office)
- Client or owner
- Number of contractors bidding on the project
- Profit or loss potential
- Construction organization's experience in this type of project
- Current and projected economy and business climate
- Time and money required to produce the bid
- Competition and their current workload
- Labor and subcontractor availability
- Required completion date of the project
- Completeness and quality of architectural or engineering bid package
- Current workload or backlog of projects
- Potential risk factors: weather, underground conditions, hazardous materials, labor, unions, materials and equipment availability
- Contract conditions and requirements
- Inclusion of liquidated damages or penalty clauses
- Stipulated payment terms

These factors will influence a contractor's decision to spend valuable resources and money on preparing a bid. Many contracting organizations bid on more than one project at once. The contractor should be selective in the projects he or she pursues; bidding efforts should be focused and optimized.

The Bidding Group

In most construction organizations, one of three individuals will be responsible for producing the bid or proposal: the business manager, the chief estimator (or estimating manager), or the proposals manager. The person who prepares the bid must be fully experienced and knowledgeable about the total construction and bidding process. He or she should be aware of all the requirements and details contained within the owner's request for bid.

A contracting organization's bidding group usually performs the activities required in the bidding process. The bidding group is directed by one of the individuals listed above. The bidding group usually is comprised of two or three estimating personnel and certain individuals from the purchasing, engineering, and project management groups.

Throughout the bidding process, the owner, or the owner's architect or engineer, is often fine tuning and revising drawings, specifications, and bid documents. This is usually accomplished by addenda that are forwarded to all bidders. This, along with decisions the contractor has to make regarding profit, markups, quantity takeoffs, site visits, vendor pricing, management reviews, and other requirements make the bidding process an intense, difficult, and time-consuming activity for a contractor.

The Bidders List

The purchasing department should maintain a list of vendors, suppliers, and subcontractors who have been qualified as potential suppliers of materials, fabricated equipment, and services. It is the purchasing department's task to approve and qualify these bidders.

The following information should be obtained from the potential bidders prior to including them on the approved bidders list.

- Management ability: names and experience
- Manpower capacity and qualifications
- Financial resources
- Bank references
- Past performance in meeting similar design, quality and delivery requirements
- Labor contract's expiration date (if union labor is used)
- List of current disputes or litigation actions
- Present and projected workload
- Installation capabilities
- Current insurance coverage and limits
- Bonding capacity
- References and contacts of previous customers

The Invitation to Bid

One of the most important elements in purchasing is the *invitation to bid* or the *request for quotation*. The invitation could be issued by an owner to a general contractor, by a general contractor to a subcontractor, or by an engineering procurement and construction management organization to a potential contractor or vendor. The invitation to bid specifically identifies the items that are to be purchased and the terms and conditions under which the sale will be performed.

The invitation to bid should consist of the following elements:

- Invitation to bid letter
- Equipment and project specifications and applicable drawings
- Instructions to bidders
- The closing date, time, and location when bid is to be submitted
- Preprinted form requesting bidder to indicate within 72 hours of receipt of invitation to bid whether a valid bid will be forthcoming
- Commercial bid form
- Data submittal requirements
- Copy of general terms and conditions; project purchase order or subcontract documents
- Shipping, QA, and start-up requirements
- Special conditions and site factors
- Insurance or bonding requirements
- A clause reserving the contractor's right to accept or reject the proposal in whole or in part

The Invitation to Bid Letter: The invitation to bid letter is the most important document in the invitation to bid package. It is used to outline general purchasing information and the actual requirements of the invitation to bid. In certain cases the invitation to bid can be faxed from the contractor to the subcontractor or vendor; some organizations also use local newspapers and trade magazines to display their invitations to bid.

The bid letter should include the following items:

- Project, name, location, and project number
- Requisition number or bid package number

- Closing date for bids
- Required delivery date of materials or equipment at the project site
- Summarized scope of work description
- Any special requirements or conditions
- Contact names that the bidding organization can call for technical and commercial clarification
- Payment terms
- Evaluation method and procedure

Progress payments should be avoided whenever possible. Payment should be made when equipment is delivered to the site; however, many vendors demand some kind of progress payment. If progress payments cannot be avoided, payment should be keyed to milestone events such as commencement of work, delivery of vendor drawings for approval, or increments of 25%, 50%, and 75% completion of fabrication, and so forth.

Figure 5.3 is an example of an invitation to bid letter that can be utilized or modified for future applications. This particular sample is from an HVAC subcontractor to an air handler manufacturer.

Bid Preparation

The following is a list of tasks that contracting organizations usually perform when preparing bids or proposals. The list may be modified for individual bidding activities.

A. Contractor receives request for bid from owner, or owner's architect or engineer.

B. Contractor reviews plans, specifications, and bid documents, noting:
   - Type of construction (steel, concrete, wood)
   - Location of construction project
   - Special or unusual features
   - Completeness of drawings and specifications
   - Type of contract (lump sum fixed price, cost reimbursable, or negotiated)
   - Current workload

C. Contractor makes decision to bid on construction project:
   - Determines whether good profit will be realized
   - Ensures that project falls within his or her area of expertise
   - Considers size of construction project, small or large
   - Considers client or owner
   - Estimates workload of competition
   - Notes number of bidders on bid list

D. Contractor attends pre-bid conference:
   - Asks relevant questions
   - Observes competition and competition's interest in project

E. Contractor performs detailed analysis of drawings, specifications, and general requirements:
   - Compiles list of work to be completed
   - Performs detailed quantity takeoff
   - Determines which party will perform each element of work
   - Produces purchasing responsibility matrix for equipment and commodities/bulk materials

From: XYZ (HVAC contractor), Inc. (date)

To: ABC (air handlers), Inc.

Attention: ______________________________

SUBJECT: MAY FOOD CORPORATION
New Manufacturing Facility Project
Bid Package No. 123-456

Dear Sir or Madam:

Your organization has been selected as a possible supplier for the above mentioned construction project, and is hereby invited to submit a bid on subject equipment. Your bid must arrive at the above address by ___(insert due date)___.

Please read the Instructions to Bidders carefully. Failure to respond to all the information requested, or failure to provide an explanation for omitting relevant information, or to submit bid on or before the due date will be grounds for considering your bid nonresponsive. Please use the attached Bid Form (Attachment I) to submit your bid. Be sure to fill it out completely; items not completed will be considered part of the bid price.

Instructions to Bidders

- The scope of work for this Invitation to Bid includes the following: Furnish and supply seven (7) Air Handler Units complete with accessories in accordance with the attached specification no. 785-15-005 and drawing nos. 119, 120, 121, and 122, all designated revision "A".
- All required accessories identified in the specification and on the drawings should be included in the bid price; failure to do so may deem your bid nonresponsive and unacceptable.
- The F.O.B. point shall be at the construction job site.
- The cost of freight from point of manufacture must be included in bid price.
- All applicable taxes, including sales or use taxes, must be included in bid price.
- Upon award of purchase order two (2) mylar drawings for each shop drawing and certified drawing submitted for approval/record will be required. Where shop drawings are not applicable, (e.g., catalog cuts, fabrication sketches, O & M Manuals), two copies of each will be required of successful bidder.
- A 10% retention on payment will be withheld until start-up and acceptance of equipment.
- Any questions of a technical nature should be addressed to Mr. John Smith, telephone 555-1293. Questions on commercial subjects should be addressed to Mrs. Sharon Jones, telephone 555-1364.

Sample Invitation to Bid Letter

Figure 5.3a

- Any exceptions to this Invitation to Bid, whether technical or commercial, should be clearly shown on the Bid Form.
- Suppliers should submit two (2) priced copies of their bid.
- All warranties and guarantees will be in effect for one (1) year from start-up and acceptance of equipment.
- Air handlers will be required on construction site on the following date: (insert date required).
- Bids shall be open for acceptance for 90 days after bid due date.

Bids will be privately opened, with all contents considered privileged information. This information will not be revealed to or discussed with anyone other than the personnel evaluating the bids. We reserve the right to reject any and all bids.

We request that you return one copy of Attachment II of this Invitation to Bid within 72 hours indicating, in the Acknowledgement Block, your intention to either bid or decline this invitation.

Yours truly,

Sharon Jones
Assistant Purchasing Manager
XYZ (HVAC contractor), Inc.

Enclosures: Attachment I (Bid Form)
Attachment II (Intention to Bid or Decline)
Specification No. 785-15-005
Drawing Nos. 119, 120, 121, & 122, all designated revision "A"

Sample Invitation to Bid Letter (continued)

Figure 5.3b

- Reviews all sections of the request to ensure that no work element is excluded

F. Contractor notifies subcontractors and suppliers and obtains bids:
- Solicits bids from subcontractors, vendors, and suppliers
- Obtains or makes copies of drawings and specifications for subcontractors, vendors, and suppliers

G. Contractor conducts site visit:
- Determines site conditions
- Evaluates access to site
- Lists demolition items, if any
- Determines underground conditions — soil, water, and rock
- Evaluates need for temporary roads, laydown areas, offices, electricity, water, and gas
- Determines obstructions, overhead or underground cables or pipelines
- Determines permit requirements and cost
- Takes photographs, talks with individuals knowledgeable with local labor conditions
- Prepares list of questions for items that are unknown or unclear

H. Contractor compiles bid:
- Selects most competitive pricing from subcontractors, vendors, and suppliers
- Estimates man-hours and prices out labor costs, including all overheads for own labor force
- Totals all direct and indirect costs required on project, including general conditions
- Adds markups and profit percentages to bid
- Adds costs related to commercial terms of bid, payment terms, retainage, insurance, and so on
- Reviews with senior management, assesses risk items and potential profit margins
- Compiles "basis of estimate," lists exclusions, inclusions, and any qualifications or alternates, lists drawings and specifications on which bid is based, including all revisions and addenda

I. Contractor submits and delivers bid:
- Ensures that bid is delivered to correct place, on schedule and to correct person, and that bid security requirements are maintained

J. Contractor awaits award decision:
- Refines and develops execution strategy if successful
- Reviews and critiques the bidding process if unsuccessful. Compiles a list of errors, mistakes, etc. Ensures that these errors and mistakes won't happen on future bidding situations.

Receipt of Bids From Subcontractors, Vendors, and Suppliers

Depending on the administrative arrangements of the construction project, bids may be directed to either the Manager of Purchasing or the Project Manager. Bids should remain unopened until the bidding period expires.

In order to quantify the number of bids that will be received, the purchasing department should contact each bidder and confirm receipt of the bid package and address any initial comments or questions regarding scope of work. This activity should happen within 72 hours of the mailing of the bid package. Attachment II is usually sent by the bidders by mail or by fax to the purchasing department.

Sealed and Unsealed Bids

Each organization should determine the procedure to be followed when receiving bids, based on the estimated value of the requisition. For example, a requisition with an estimate value exceeding $10,000 might follow a sealed bid procedure; requisitions under $10,000 might follow an unsealed bid procedure. Either procedure would require that all bids be time- and date-stamped and logged into the requisition summary form. Unsealed bids would be stamped on the Attachment I bid form; sealed bids would be stamped on the outside of the envelope.

Ideally, procedures for sealed bids should follow these basic guidelines:

- All bids with a value exceeding $10,000 should be submitted in a sealed inner envelope contained within a sealed outer envelope.
- The sealed inner envelope should contain two copies of the bid. The outside of the inner envelope should be labeled as follows:
  1. Project name/number (as shown on bid documents)
  2. Requisition number (as shown on bid documents)
  3. Submitted by (name and address of bidder)
  4. Bid due date and time
- The outer envelope should be sealed, addressed, and referenced in accordance with the instructions to bidders given in the invitation to bid.
- The incoming sealed envelopes from bidders should be delivered unopened to the purchasing department, where they should remain unopened until the bid due time and date.
- The opening of all sealed bids should be properly witnessed by the appropriate senior project or purchasing management. The witnessing and review should include the identification and logging in of each bid received, including the time and date and the value of each bid. All bid documents should be filed for future record and auditing requirements.
- Bids received after the due date should be returned unopened to the bidders together with a letter explaining why the bid is being returned.

The procedure for unsealed bids should endeavor to follow the same principles and procedures where possible. The basic approach for unsealed bids is typically less formal and stringent in light of the lower requisition value. Further details regarding the receipt of subcontractor bids are discussed later in this chapter.

After all bids have been received and logged in, the bid review and evaluation cycle can be started.

Bid Review, Analysis, and Recommendation

Immediately after the bid opening procedure, the purchasing department should determine which bids are the most commercially attractive. An evaluation form should identify and rank bidders in order of most competitive bid value. This form should be forwarded to the engineering department or project management along with any relevant bidders' technical documentation. This allows the appropriate personnel to commence the technical review and evaluation of the bids.

The price transmittal form serves as a preliminary basis for determining which supplier, vendor, or subcontractor bids should be considered. Typically, if six to nine bids are received, the lowest three or four of these bids should be considered.

All related communications, questions, and correspondence with the suppliers, vendors, and subcontractors during the bid and evaluation cycle should in most cases be performed by the purchasing department. This will ensure that all bidders receive the same information regarding any clarifications or changes during the evaluation phase. The purchasing department typically has the most practical understanding of the current questions or problems related to the specific bid package.

The Bid Tabulation Summary: The bid tabulation summary form (or "bid tab") lists all the items contained in the bids received. The information contained in this document forms the basis of the evaluation of bids and the recommendation of award. Any exceptions, qualifications, or missing scope items should be incorporated into the bid tab, which should be clear and easy to read.

The following is list of items typically found in the bid tabulation summary:

- Deviation from specification drawings and instructions to bidders
- Freight and transportation costs
- Packing costs or requirements
- Escalation values or formula basis, if applicable
- Delivery dates
- Warranty and performance guarantees
- Payment terms
- Unit price listing and details for additional or deleted work
- Spare part costs
- Vendor assistance charges if required during start up of equipment
- Shop tests during manufacture of equipment
- Bonding capacity, if applicable
- Taxes, included or excluded

An example of the tabulation summary form is shown in Figure 5.4. Once the bid tab has been compiled, the purchasing department and project management should sign and approve the recommendation.

Final Bid Approval Form: A final bid approval form is also an option. This form allows input from senior management and owner management, if this is required. Such approval gives the purchasing department the authority to formulate a purchase order and/or subcontract agreement.

Figure 5.5 is an example of a final bid approval form, which can be completed either after the bid tab is submitted or whenever an owner requests to be part of the approval process.

Bid Tabulation Summary

Project No. ____________
Project Name ____________
Project Location ____________
Client ____________
Date ____________ Rev. ______
Page ____________ of ______
Prepared By ____________

Description/ Listing of Items	Bidder ____ Address ____ Bid No. ____	Bidder ____ Address ____ Bid No. ____	Bidder ____ Address ____ Bid No. ____	Bidder ____ Address ____ Bid No. ____	Bidder ____ Address ____ Bid No. ____
Base Bid:					
Freight Costs:					
Packing Costs:					
Taxes:					
Escalation:					
Bonds:					
Other:					
Other:					
Other:					
Total Cost					
Meets Spec:	☐ Yes ☐ No	☐ Yes ☐ No	☐ Yes ☐ No	☐ Yes ☐ No	☐ Yes ☐ No
Bid Expir. Date:					
F.O.B. Point:					
Warranty Period:					
Other:					

Comments: ____________

Approvals/Recommendations

Award to Bid No. ____________
Bid Value: $ ____________
Signed by Purchasing ____________ Date ______
Signed by Engineering ____________ Date ______
Project Manager ____________ Date ______

Figure 5.4

CLIENT: **<u>Johnson Carpet Manufacturing Co.</u>**

3 Air Handling Units
Project No. 498-987

(insert date)

Recommended Supplier: Jackson Air Handling Corporation

Bid Price:	$191,764.00	FOB Job site
Fabrication Start Date:	Immediately	
Delivery:	Within 12 weeks	
Estimate Budget:	$210,000.00	

COMMENTS:

JAHC accepted all of the negotiable terms and conditions including a payment term of 90% upon delivery and 10% retainage to be paid after acceptance of equipment. JAHC was lowest acceptable bidder and can meet field delivery need dates.

	Signature	Date
Preparer		
Purchasing Manager		
Project Manager		
Operations Manager		
Client (optional)		

Sample Final Bid Approval Form

Figure 5.5

Bid Negotiation: For most purchases, some degree of negotiation will be necessary prior to placement of the order. The objective of any negotiations should be to obtain the maximum value for the least amount of money while still allowing the bidder to realize a fair and reasonable profit on the equipment, goods, or services he or she is providing.

Recommendation of Awards: On construction projects with large value purchase orders and subcontracts, pre-award meetings with the most competitive bidders may be required. The objective of the pre-award meeting is to ensure that bidders have a full understanding of the complete scope of work and the technical, quality, commercial, and schedule requirements of the purchase order or subcontract. The meeting should also confirm the bidder's ability to meet the required delivery date of equipment, shop drawings, and other necessary data. Conference notes detailing major items of discussion and agreement should be prepared and signed by the bidder before the meeting is adjourned in order to document agreements and ensure complete understanding by the bidder of all obligations and requirements related to the purchase order or subcontract.

When all relevant commercial and technical matters that could impact the selection of the most attractive and acceptable bidder have been concluded, the purchasing department compiles and presents its recommendation to the project manager. This recommendation will consist of the following items:

- The completed bid tabulation form
- A technical evaluation and recommendation memo signed by the technical evaluator
- Bid conditioning estimates and estimating basis (if required)
- Pre-award meeting notes
- The recommended bidder's name

The recommended bidder will usually be the organization offering the most appealing combination of technical acceptability, quality, delivery, field support, and, most importantly, the lowest competitive price.

The Letter of Intent

If a contract or purchase order cannot be completed in a timely manner, the purchasing department will issue a letter of intent to inform the successful bidder that he or she has been awarded the contract. The purpose of the letter of intent is to allow the successful bidder to commence the work or to order any long lead delivery components associated with the contract prior to receiving and executing the detailed contract. Figure 5.6 is an example of a letter of intent.

Notice to Unsuccessful Bidders

It is a good business practice to keep your bidders apprised of the bidding situation and to inform them as soon as possible if they have been unsuccessful. Figure 5.7 is an example of a notice to unsuccessful bidders.

SODER CONSTRUCTION AND DEVELOPMENT CORPORATION

(date)
____

Mr. David Davis
Jackson Air Handling Corporation
123 Main St.
Dallas, TX 75309

Subject: Letter of Intent – Project No. 498-987. Three A/H Units
P.O. 071
Client: Johnson Carpet Co.

Dear Mr. Davis:

This letter is to advise you that it is our intent to purchase from your organization 3 Air Handling Units for the sum of $191,764 F.O.B. job site at the new Johnson Carpet Co. project site.

Jackson Air Handling Corporation promised to deliver 3 Air Handler Units in accordance with specifications No. 15-683-SK9 (A) & (B). Delivery will be achieved within 12 weeks after commencement of manufacture, which will commence (insert date) . All other terms and conditions remain in place and in accordance with your proposal dated (insert date) referenced 58-2001 revision "A".

An original and duplicate copy of this letter is enclosed. If this letter is satisfactory to you please indicate your acceptance by signing the original copy and returning it to us immediately. A formal contract will be forwarded to you within the next two weeks.

Very truly yours,

P.H. Mussen
Purchasing Manager

Accepted by ____________________________________
Signature

Jackson Air Handling Corporation

____________________________________ ____________
Name & Title Date

Sample Letter of Intent

Figure 5.6

SODER CONSTRUCTION AND DEVELOPMENT CORPORATION

Subject: Project No. 498-987
P.O. 071 – 3 A/H Units
Client: Johnson Carpet Co.

Dear Sir or Madam:

Following a detailed commercial and technical evaluation of your bid on the above subject project, we have selected another bidder's proposal. We would like to express our sincere appreciation for your effort and cooperation. If we have any future bidding requirements, we will contact your organization.

Very truly yours,

P.H. Mussen
Purchasing Manager

Sample Notice to Unsuccessful Bidders

Figure 5.7

Subcontracts

As mentioned earlier, the normal practice in construction is for the general contractor to subcontract certain work items to specialty contractors, who are usually referred to as subcontractors. The subcontractor performs certain specified portions of the construction project.

The written subcontract agreement defines the rights and obligations of both the contractor and the subcontractor. The contractor should endeavor to pass on to the subcontractor the same terms, conditions, and contract clauses that are part of the prime contract between the owner and the contractor. In this way, the contractor ensures that the project risk is equally distributed among the construction team members (i.e., contractor and subcontractors).

The Subcontracting Process

The following list reflects the steps involved in the subcontracting procurement cycle:

A. Contractor receives request for proposal from owner:
- Reviews bid package
- Makes decision to bid

B. Contractor commences subcontract planning (during bid phase):
- Develops subcontractor strategy
- Defines scope of work to be subcontracted
- Prequalifies subcontractors; issues questionnaire to subcontractors
- Issues preliminary bid package to subcontractors
- Obtains lump sum bids from subcontractors
- Performs technical and commercial evaluation of subcontractor bids

C. Contractor compiles and issues bid to owner:
- Includes most favorable bids from subcontractors
- Includes own direct and indirect costs
- Submits bid

D. Owner awards construction contract

E. Contractor reconfirms that subcontractors' bids contained within contractor's proposal are correct and valid

F. Contractor conducts pre-award meeting with subcontractors

G. Contractor negotiates and awards subcontracts to subcontractors

Qualifying Subcontractors

It is extremely important that contractors use highly qualified and reputable subcontractors. The attributes that subcontractors should possess include:

- Experience and technical knowledge of work to be performed
- Competent management and supervision
- Financial strength and stability
- Adequate manpower resources
- Properly maintained equipment, facilities, and plant
- Adequate bonding capacity

The normal practice is for the purchasing department to conduct a thorough investigation on any potential subcontractor. It is very important that past performance on similar construction projects is analyzed in detail. References and credit reports should be carefully researched and evaluated.

Various organizations, such as the American Institute of Architects and the Associated General Contractors of America, produce prequalification questionnaire forms that are used for qualifying subcontractors. Figure 5.8 is an example of such a questionnaire form.

The Subcontract Document Package

The preparation of a subcontract document package is an important element of subcontracting. It is very important that the contractor include the entire scope of work to be performed and any additional items that he or she expects the subcontractor to bid on and eventually fulfill. Any deficiencies or mistakes in the subcontract package can result in contractual, financial, and legal disputes between the contractor and the subcontractor.

Some contractors design their subcontract packages according to the format established in the construction contract document package. In this way the contractor passes onto the subcontractor some of the risk elements of the construction contract even though there is no direct contractual relationship between the subcontractor and the owner. Many contractors use preprinted standard agreements that can be modified for each new subcontract. Certain clauses can be modified or deleted as the situation warrants. Figure 5.9 is an example of a typical subcontract document package, which a contractor would send out to several prequalified specialist contractors.

Subcontractor Selection

The goal of a contractor who is in a competitive bidding situation is to produce a low winning bid that will be accepted by the owner but will also allow a fair and reasonable profit margin. Achieving this goal requires careful planning and detailed analysis of subcontractor bids and eventual award.

Major subcontractors historically perform the electrical and mechanical work. On industrial or manufacturing construction projects, major subcontractors include the electrical and mechanical subcontractors as well as piping/equipment installation subcontractors and instrumentation subcontractors. Minor subcontractors usually include the painting, insulation, and landscaping subcontractors.

Plan and Specification Bid: Many contractors, especially in the commercial segment of construction, use the plan and specification approach for obtaining subcontractor bids. When bidding to an owner, the contractor will inform the local subcontractor community that he is interested in obtaining subcontractor bids on a specific construction project. The contractor will inform the subcontractors that the bid package documents are available for review at the contractor's home office or branch office. The subcontractor can usually have a copy of these documents for a nominal fee (usually $50 to $100).

Many subcontractors are initially interested in bidding, but of the total number of interested subcontractors, less than 50% will actually submit valid bids to the contractor. Many will not like the general conditions, terms of the contract, or the bid due date on "plan and spec" work; some may not be capable of performing the work.

Subcontractor Questionnaire Data Form

Project No. ____________________
Project Name ____________________
Project Location ____________________
Date ____________________
Prepared By ____________________
Page _____ of ____________________

Organization

Name of Company ____________________

Street Address ____________________

City and State ____________________ Zip Code ____________________

Telephone No. ____________________ FAX No. ____________________

1. Indicate type of business organization.

   ☐ **Corporation**. List names of officers.

   President ____________________ Secretary ____________________

   General Manager ____________________ Treasurer ____________________

   Place of Incorporation ____________________ Date ____________________

   ☐ **Partnership**. List names of partners.

   ____

   ____

   ____

   ____

   ☐ **Sole Owner**. Name ____________________

2. Names of owners (stockholders holding over 10% of stock):

   ____________________ ____________________

   ____________________ ____________________

   ____________________ ____________________

3. Principals of company (officers' names, titles, qualifications, experience and years):

   ____

   ____

   ____

   ____

4. Subsidiaries (indicate whether wholly owned or percent controlled):

   ____

   ____

5. Number of years in business under your present name: ____________ years.

6. The company is ____________% minority owned.

7. Indicate the number of permanent employees currently on payroll:

   Management ____________

   Engineers ____________

   Draftsmen ____________

   Office admin. staff ____________

   Field supervisors ____________

   Field labor force ____________

   Total ____________

Figure 5.8a

Subcontractor Questionnaire Data Form (cont'd)

Project No. ______________________
Project Name______________________
Project Location ______________________
Date______________________
Prepared By______________________
Page_____ of______________________

CLASSIFICATION OR TYPE OF WORK PERFORMED

1. Check type of construction work your company employees perform.

- ☐ Architectural work
- ☐ Carpentry work
- ☐ Concrete work
- ☐ Conveyors/Elevators
- ☐ Demolition/Relocations
- ☐ Electrical work
- ☐ Excavation work
- ☐ Fencing work
- ☐ Fire protection work
- ☐ Glass/glazing work
- ☐ Hazardous Waste Removal
- ☐ H.V.A.C. work
- ☐ Inspection & testing work
- ☐ Insulation/heat tracing work
- ☐ Lab equipment installation work
- ☐ Landscaping work
- ☐ Masonry/brickwork
- ☐ Mechanical work
- ☐ Millwright work
- ☐ Painting work
- ☐ Paving work
- ☐ Pile driving work
- ☐ Piping systems
- ☐ Plumbing work
- ☐ Roofing work
- ☐ Sheet metal work
- ☐ Site work
- ☐ Sprinkler work
- ☐ Structural steel work
- ☐ Tunnelling work
- ☐ Others (List)
  ____
  ____

2. Percent of work performed as a general contractor ____________%
3. Percent of work performed as a subcontractor ____________%
4. List type of work usually subcontracted to others ______________________
____
____

WORK HISTORY

1. List the important projects completed by your organization within the last five years including the contract value.

Name of Client	Person to Contact	Project Title and Scope of Work	Contract Value	Year Work Performed

FINANCIAL

1. Submit last three annual financial reports and current profit and loss statement (audited report preferred).
2. a. What is the maximum dollar value of a project you believe your company is capable of handling?
   $______________________
   b. Over what period of time? ______________________
3. Average annual dollar volume of work for the past five years $______________________
4. Is there any litigation now in progress or pending with clients, subcontractors, or suppliers? ☐ Yes ☐ No
   If yes, give details. ______________________
   ____
   ____

Figure 5.8b

Subcontractor Questionnaire Data Form (cont'd)

Project No. ______
Project Name ______
Project Location ______
Date ______
Prepared By ______
Page ____ of ______

5. a. Do you have an established bonding company? ☐ Yes ☐ No
   b. If yes, name of bonding company ______
   Address ______
   Contact ______ Phone No. ______ Bonding Capacity $ ______
6. Indicate banking references ______

CONSTRUCTION EQUIPMENT AND MACHINERY

1. List owned construction equipment with capacity, age, type and attachments.

LABOR AND UNION AFFILIATIONS

1. Does your organization perform work as an open shop? ☐ Yes ☐ No Closed shop? ☐ Yes ☐ No
   If both, is work performed under same name? ☐ Yes ☐ No If different names, please list both.

2. Do you have any union national agreements? ☐ Yes ☐ No If yes, with which crafts?

3. If you are signator on local agreements, indicate the following:

Craft and Local	Holder of your bargaining rights

SAFETY AND INSURANCE INFORMATION

1. Person responsible for safety program ______
   Title ______ Phone No. ______
2. Person to be contacted for matters involving insurance ______
   Phone No. ______ FAX No. ______
3. Insurance agent's name and address ______

4.	Last Year	Previous 5 Yrs.
a. Number of lost workday cases (injuries involving days away from work)		
b. Number of cases with medical treatment only		
c. Number of fatalities		

This statement was completed by:

Name ______ Title ______

Signature ______ Date ______

Figure 5.8c

(date)

XYZ Contractors, Inc.
123 Main Street
Anytown, NY 10500

ABC Site Work Construction Co.
321 Wilson Street
Anytown, NY 10500

Subject: New Building for New Age Research Corp.
Anytown, NY
Project No. 152

Request for Proposal: Site Work
SC-001

Quotation Due Date: (insert date)

Dear Sir/Madam:

Your company has been selected as a possible subcontractor for the above project, and is hereby invited to submit a lump sum proposal on the subject work.

Please read all documents carefully. Your attention is directed to the enclosed Intention to Bid Confirmation Form, which must be returned to the writer within 5 days of receiving this package.

We thank you for your interest in this important project.

XYZ Contractors, Inc.

John Smith
Purchasing Agent

Enclosure No. 1. Intention to Bid Form
2. Instructions to Bidders
3. Bid Form
4. Attachments to Bid
5. Unit Prices
6. Scope of Work
7. Subcontract Agreement
8. Project Specification
9. Contract Drawings

Sample Subcontract Document Package

Figure 5.9a

Enclosure No. 1 INTENTION TO BID FORM

(RETURN THIS PAGE WITHIN 5 DAYS OF RECEIPT OF BID PACKAGE TO:)

XYZ Contractors, Inc.
123 Main Street
Anytown, NY 10500

ATTN: Mr. John Smith, Purchasing Agent

Subject: New Building for New Age Research Corp.
Anytown, NY
Project No. 152

Subcontract Inquiry: Site Work
SC-001

Dear Sir/Madam:

We hereby acknowledge receipt of the complete set of bidding documents for the above referenced project work, and confirm that:

(Check appropriate box)

☐ We do intend to submit a proposal for this work. We understand that this proposal will be prepared by us at no cost or obligation to XYZ Contractors, Inc. or New Age Research Corp.

☐ We do not intend to bid on this work, and are hereby returning the bidding documents received. The reason(s) we decline to offer a proposal are as follows:

____

____

____

Yours sincerely,

Name	Signature
Firm	Title
	Date

Sample Subcontract Document Package (continued)

Figure 5.9b

Enlosure No. 2 INSTRUCTIONS TO BIDDERS
On the new building for
New Age Research Corp.
Anytown, NY

- Sealed bids for the work described within this document will be received at the following address on or before January 17, 19XX, 3:00 p.m., and will be privately opened.

XYZ Contractors, Inc.
123 Main Street
Anytown, NY 10500

ATTN: Mr. John Smith

- The bids, based on the Subcontract Bidding Documents, shall be held as responsible lump sum bids made with full knowledge of conditions and requirements of the work to be performed. Bidders are directed to visit above site prior to the time of submitting the bid and thoroughly inspect the conditions under which the Contract is to be executed. No consideration will be given to requests for additions to the contract lump sum amount for work resulting from existing conditions or restrictions on the use of the premises which could have been determined during the bid period, whether or not that work is indicated or described on the contract drawings and specifications.
- Bids are to be made on the forms provided and enclosed in an envelope, sealed and endorsed with the name of the work as follows: Confidential Bid for New Building for New Age Research Corp., reference 152, SC 001.
- Bids shall be open for acceptance for 90 days after bid due date.
- Bids not submitted on the forms provided and in accordance with the instructions contained herein shall be considered informal, and at the option of XYZ Contractors, Inc., may be rejected.
- The Subcontract Bidding Document Package includes:

  Intention to Bid Confirmation Form
  Instructions to Bidders
  Bid Form
  Attachments to Bid
  Unit Prices Form
  Scope of Work
  Sample Subcontractor Agreement
  Project Specifications
  Contract Drawings

- Bids must be submitted on the prescribed Bid Form in duplicate. All blank spaces for prices must be filled in, in ink or typewritten, and total amounts indicated in both words and figures. Lump Sum Bid is to be firm through completion of field construction with no claims for material or labor cost escalation.
- Bidder shall submit his bid in accordance with this inquiry and attachments. Alternate proposals may be submitted provided they are labeled "ALTERNATE" and all deviations from the specification are individually indicated and described.
- Bidder must be licensed to perform business as a contractor in the State where job site is located if State and/or local municipal laws require such licensing.

Sample Subcontract Document Package (continued)

Figure 5.9c

- All work and/or services, including delivery to job site, to be provided by the subcontractor shall be performed by utilizing respective Trade Union labor. The subcontractor is directed that the performance of all work under this subcontract shall be in accordance with current local trade line jurisdiction practice.
- Any additional or supplemental instructions will be in the form of written Addenda to the scope of work. All addenda so issued shall become part of the contract documents and shall be acknowledged in the bidder's proposal in the space provided therefore.
- The attention of each bidder is directed to the requirements for submission of a proposed duration to perform the work within a consecutive calendar day period allowing for the reasonable coordination and scheduling of work by other subcontractors consistent with standard construction industry practice. Each bidder shall complete and submit a projected manpower loading schedule, indicating respective trades assigned to perform the work. Manpower loading schedules are not required to be submitted with the bidders proposal, but will be required prior to bid review, if so directed by XYZ Contractors, Inc.
- Bidders shall include in their bids any and all amounts required for all applicable taxes, including sales or use taxes as applicable to complete their work.
- Bidders shall include in their proposals any and all costs necessary to obtain required permits to perform their work.
- XYZ Contractors, Inc. reserves the right to accept or reject any or all bids and further reserves the right to award the work to the bidder who XYZ Contractors, Inc. deems is best suited to perform the work.
- The basis for award, subject to the conditions stipulated above, shall be the firm lump sum base bid amount for all of the work required and described by the contract documents package.
- Each subcontract bidder is required to identify by name any subcontractors that he proposes to use for the work should he be awarded the contract. In addition to the above requirement, each subcontract bidder shall identify the lump sum dollar amount to be allocated to each of the respective named subcontractors. The above information shall be shown in the space(s) provided on the Attachments to Bid Form. Each bidder shall provide adequate background information and details in the form of references, past work performed, etc., to establish the specialty subcontractor's acceptability to the XYZ Contractors, Inc., if so requested.
- Bidders who require access to project site and premises during bid period should contact:

  Mr. John Smith
  Purchasing Agent
  XYZ Contractors, Inc.
  123 Main Street
  Anytown, NY 10500
- Unit prices must be submitted with the bid. Bidders are directed to complete the bid form sections listing markup percentages to be applied to changes in the Work.

Sample Subcontract Document Package (continued)

Figure 5.9d

- All lump sum bids shall be based on the materials or equipment specified within this bidding document package. Any proposed deviation from the requirements of the contract shall be listed separately on the Bid Form, and any change in the lump sum amount identified. Within seven days from the bid due date, each bidder shall prepare a list of specific equipment or material manufacturers proposed for use on the project, if so requested by XYZ Contractors, Inc.
- For questions regarding information contained in this inquiry, please contact the following:

  Commercial questions: Mr. John Smith, Purchasing Agent

  Technical questions: Mr. Robert Jones, Engineer
- Bid, payment, and performance bonds, if requested, shall be furnished by the subcontractor and issued by a surety company acceptable to XYZ Contractors, Inc. The cost of such bonds, if required, shall be reimbursed separately from the proposed lump sum amount.

BID DUE DATE: January 17, 19XX
PRE-BID MEETING: January 8, 19XX
TENTATIVE CONTRACT VERBAL AWARD DATE: February 15, 19XX
START IN FIELD DATE: March 1, 19XX
COMPLETE CONSTRUCTION DATE: June 15, 19XX

Sample Subcontract Document Package (continued)

Figure 5.9e

Enclosure No. 3 BID FORM

XYZ Contractors, Inc.
123 Main Street
Anytown, NY 10500

Attn: Mr. John Smith, Purchasing Agent

Subject: New Building for New Age Research Corp.
Anytown, NY
Project No. 152
Site Work SC 001

We the undersigned company ______________________________
having carefully reviewed the Subcontract Bidding Document Package for the above subject Project, and being satisfied that we fully understand them, upon written notice of award of contract agree to provide all supervision, labor, materials, construction equipment, plant, facilities, services, safety and insurance. Our bid also includes all taxes, permit costs, fees, and other costs necessary or required in the Contract documents, to accomplish in a safe, timely and workmanlike manner the Work described in the Subcontract Bidding Document Package for the following firm lump sum base bid:

____
$ (written amount)

The complete scope of work covered by the above firm lump sum is included in the attached breakdown in the "Attachments to Bid" form for the purpose of bid evaluation by XYZ Contractors, Inc.

Company: ______________________________

By: ______________________________
(Signature)

Title: ______________________________

Date: ______________________________

Witnessed by: ______________________________
(Signature)

Title: ______________________________

Name: ______________________________

Date: ______________________________

Sample Subcontract Document Package (continued)

Figure 5.9f

Enclosure No. 4 ATTACHMENTS TO BID

SUBJECT: New Building for New Age Research Corp.
Anytown, NY 10500

SUBCONTRACT INQUIRY: Site Work SC 001

Bid Due Date: January 17, 19XX

We hereby submit the following information as directed by the Subcontract Bidding Document Package for the subject work, which is included in and a part of our bid to perform the work:

Item Description	Quantity	Unit	Material Cost	Labor Cost	Equipment Cost	Total Cost
Demolition Work		L.S.				
Site Grading		L.S.				
Borrow Fill		C.Y.				
Paving		S.F.				
Topsoil		C.Y.				
Final Grading		L.S.				

Total Firm Lump Sum Price $ ______________________________

The cost to provide the insurance coverage in the amounts required by the Subcontract Bidding Document Package, which is included in the above firm lump sum price, is $____________.

The cost to provide the permits required by the Subcontract Bidding Document Package, which is included in the above firm lump sum price, is $____________.

The cost to provide the local and state taxes required by the Subcontract Bidding Document Package, which is included in the above firm lump sum price, is $____________.

Bid, payment, and performance bonds for this project, if required by XYZ Contractors, Inc., are not included in the above lump sum price. The cost of bonds, if required, is $____________.

We propose that for changes in scope to this contract, amounts billed for additional work performed will be marked up as follows:

Labor Rate				
	Bare Cost	Overhead	Profit	Total
Materials	N/A			
Labor				
Subcontractor				
Third Party Costs				
Equipment Rental	N/A			

Sample Subcontract Document Package (continued)

Figure 5.9g

To perform the work of this subcontract, making allowance for the normal coordination and progress of the work of other subcontracts, equipment delivery, etc., we anticipate the following number of consecutive working days will be required to complete the work of this subcontract: ________________ days.

To perform the subject work, we propose to issue the following sub-subcontracts, and use the named specialty sub-subcontractors to perform portions of the Work we do not self-perform:

Work Element	Sub-subcontractor	Name and Address	Value of Work

This lump sum bid, including unit prices, markups, alternates, etc., will remain in effect for ninety (90) days from the due date of bid. Addenda to the construction documents are included in the above price and receipt thereof is acknowledged as follows:

Addendum & Description	Dated
No. ____________________	____________
No. ____________________	____________
No. ____________________	____________

Provide a separate sheet itemizing any and all of the following items applicable to your bid or performance of the work.

- Exceptions to Bid
- Alternates & Effect on Bid
- Qualification Statements

Company: ______________________________

By: ______________________________

(Signature)

Title: ______________________________

Date: ______________________________

Sample Subcontract Document Package (continued)

Figure 5.9b

Enclosure No. 5 UNIT PRICES

Subject: New Building for New Age Research Corp.
Anytown, NY 10500

Subcontract Inquiry: Site Work SC 001

Unit prices for additional or deleted work listed below shall include all necessary supervision, labor, material, plant equipment and taxes. The following unit prices shall apply for all additions or deletions to the subcontract package.

		Additional Work			**Deleted Work**		
Items	**Unit**	**Labor**	**Mat'l**	**Total**	**Labor**	**Mat'l**	**Total**
Borrow Fill	C.Y.						
Paving	S.F.						
Topsoil	C.Y.						
Other							

Enclosure No. 6 SCOPE OF WORK

The subcontractor shall provide all necessary materials, labor, equipment, tools, services, support and supervision of the furnishing and installing of all the site work as indicated on the contract drawings and described in the specifications, and in accordance with the Subcontract Bidding Document Package. The work shall include but not be limited to: excavation, backfilling and compaction, rough grading, final grading, soil erosion control measures, repair of existing roadway base, construction of new roadways, paving and demolition and removal of paved areas, and layout work.

Sample Subcontract Document Package (continued)

Figure 5.9i

Enclosure No. 7 SUBCONTRACT AGREEMENT

Every contractor has his or her own views and thoughts, which are usually based on experience, on what should be contained within a subcontract agreement. The subcontract agreement defines the rights and obligations of both the contractor and subcontractor. The subcontract agreement is basically an operating procedure on how the contract is to be administered and coordinated. There are certain preprinted subcontract agreements available to contractors and subcontractors. The American Institute of Architects Form A401 Contractor-Subcontractor Agreement Form (5/87) is one example and is widely used throughout the construction industry.

The following is a list of items that should be considered and possibly be expanded on and included in a subcontract agreement:

- Definitions
- Administration of the agreement
- Obligations of subcontractors
- Compensation and terms of payment
- Extra work or variations to the work
- Completion of the work
- Termination default and suspension
- Delays, extension of time
- Claims
- Arbitration
- Performance and labor/material payment bond
- Subcontractors insurance requirements
- Lien indemnification
- Sub-subcontractors
- Force majeure
- Title
- Assignments
- Warranty
- Reports
- Equal Employment Opportunity and Fair Employment Practices
- Safety
- Field office
- Storage and warehousing
- Coordination with other subcontractors
- Cleanup requirements
- Testing
- Product data
- Permits

Sample Subcontract Document Package (continued)

Figure 5.9j

Enclosure No. 8 PROJECT SPECIFICATIONS

Division & Site Work	Revision and Date	
02010	Subsurface investigation	'A' 10/10/XX
02050	Demolition	'A' 10/10/XX
02100	Site preparation	'A' 10/10/XX
02160	Excavation support systems	'A' 10/10/XX
02200	Earthwork	'A' 10/10/XX
02900	Landscaping	'A' 10/10/XX

Enclosure No. 9 CONTRACT DRAWINGS

Drawing No.	Revision and Date	
101	Site Plan	'F' 10/26/XX
102	Revised Parking Area Plan	'B' 10/26/XX
103	Grading & Drainage Plan-Sheet 1	'C' 10/26/XX
104	Grading & Drainage Plan-Sheet 3	'D' 10/26/XX
105	Soil Erosion Control Plan-Sheet 2	'C' 10/26/XX
106	Soil Erosion Control Plan-Sheet 3	'D' 10/26/XX
107	Site Details and Sections	'D' 10/26XX
108	Soil Erosion Control Details	'E' 10/26/XX

Sample Subcontract Document Package (continued)

Figure 5.9k

The contractor using plan and spec bidding will usually allow subcontractors or suppliers to fax or telephone in their bids on the bid due date. Sometimes the bids are received very close to the time when the contractor has to give his or her bid to the owner. The contractor's estimating and purchasing departments work very closely on collecting and reviewing these bids. The contractor will select the most competitive bids he or she has received. The contractor's estimating and purchasing departments will then discuss the scope of work over the phone with the most competitive bidders. The bid telephone confirmation sheet is a very important document that is used in this bidding method. One hour or in some cases 30 minutes before the bid due time, the contractor may be discussing scopes of work and pricing details with many subcontractors and suppliers. The contractor is, of course, endeavoring to get the most competitive bid he or she can get. Eventually the contractor has to select the best values he or she has obtained. The bid form is compiled, usually placed in a sealed envelope and delivered to the owner, sometimes only minutes before the bid due time. To be successful in this kind of bidding environment, the contractor must compile and maintain comprehensive and detailed reports detailing discussions and verbal agreements.

In plan and spec bidding, the contractor must ensure that the total scope of work described in the owner's plans and specifications is covered in his or her bid. Close coordination must take place on "bid day". This is the day when all the bids from subcontractors, vendors, and suppliers are analyzed for inclusion in the contractor's bid to the owner. Missing scope or duplication of scope can mean the difference between success and failure. The time constraints and the problems in dealing with the many different organizations that have submitted bids can make this form of bidding a very risky undertaking. Various industry experts have concluded that between 30% and 40% of construction contracts are performed using this particular approach.

Figure 5.10 is a sample subcontract bid telephone confirmation sheet that a contractor would complete when receiving telephone bids from subcontractors.

The Subcontractor Bid Tabulation Comparison

The process of analyzing and selecting the various subcontractor bids and incorporating the most favorable bids into the contractor's bid submission to the owner can be extremely complex and time consuming. This activity is usually performed by personnel from the contractor's purchasing and estimating departments.

The usual procedure is to complete a bid tabulation comparison sheet for all bids received from subcontractors for particular elements of the project. Certain bids will need to be conditioned or adjusted to reflect missing or incomplete items of scope that the subcontractor has excluded or qualified in his lump sum bid. All subcontractor bids must be technically and commercially reviewed to ensure that they meet the requirements of the subcontract bid package. The subcontractor bid tabulation comparison sheet should include a section in which the contractor's technical and commercial personnel indicate that the subcontractor's bid is both technically and commercially complete and acceptable.

Subcontract Bid Telephone Confirmation

Project No. ____________________
Project Name____________________
Project Location ____________________
Date____________________
Page_____ of____________________
Prepared by: ____________________
Report No. ____________________

Date: ____________________ Revision: ____________________ Time: ____________________

Owner ____________________
Subcontractor Name ____________________
Address ____________________
Telephone No. ____________________
Contact ____________________
Title ____________________
Fax No. ____________________
Bid Reference No.____________________

Scope of Work Bid: ____________________

	Yes	No
All Taxes in Bid	☐	☐
Freight	☐	☐
Plan & Spec	☐	☐
MBE/WBE Contractor	☐	☐
Shop Drawings	☐	☐
Subcontractor's Own General Conditions	☐	☐

Addenda No's. included in bid ____________________
Value: $ ____________________

Exclusions and omissions to bid: ____________________

Bid Value: $ ____________________
Alternatives: A. ____________________

$ ____________________
B. ____________________

$ ____________________
Information received by ____________________ (Date/Time) Information.
From:____________________

Note: Subcontractor must confirm bid price within 24 hours.

Comments: ____________________

Figure 5.10

Types of Subcontractor Bids

There are basically four types of subcontractor bids that a contractor receives.

Low Acceptable Bid: This is the best situation for the contractor. A subcontractor submits a complete and acceptable bid that is lower in price than bids received from other subcontract bidders, lower than the contractor's in-house budget estimate, and meets all the requirements of the subcontract bid package. In most cases, the subcontractor has been prequalified and has demonstrated that he or she possesses the ability and technical know-how to satisfactorily perform the work described.

Partial or Incomplete Bid: The contractor is faced with a problem in this situation because the subcontractor has excluded some important element of work from his or her bid. The partial or incomplete bid is almost always the lowest bid received, so to exclude this bid would be a mistake. The logical procedure is to contact the subcontractor and ask him to complete the bid or to make a budget estimate of the missing scope of work and add it to the incomplete bid. If this subcontractor is still the low bidder, the purchasing agent or bid evaluator should negotiate a satisfactory solution. The contractor could possibly perform or use other subcontractors to complete the work.

Single Bid: If only one bid is received, the only way to proceed is to compare the single subcontractor's bid with the contractor's own budget estimate. If the price is lower than this budget, it is safe to accept this single bid. If it is higher than the contractor's own budget estimate, additional bids must be obtained or the contractor may be forced to perform the work. The contractor should not inform a subcontractor that he is the only bidder; if a subcontractor knows this he will usually maintain or increase the bid price substantially because there is no competition.

Significantly Low Bids: This situation could cause a serious problem to the contractor. The subcontractor could go bankrupt, or he could pull out of the project when his work is only partially complete. Many contractors will either ignore these low bids or inform the subcontractor of the situation. Possible solutions to this situation are to request a performance bond from the subcontractor, use the next lowest acceptable bid, or for the contractor to perform the work with his own labor force.

The Subcontract Administration Process

The subcontract administration process commences with the identification by the contractor's bidding group of the work elements to be subcontracted. The subcontract administration process continues through subcontract performance and eventual close-out.

The individual who performs the subcontract administration process should have a strong background in contract administration and procurement. In certain situations, the individual handling the subcontract administration will not be working alone; he or she will need to work and communicate with other project team members.

The following is a list of activities to be considered before the commencement of the overall subcontract administration process. These activities will be performed either at the contractor's home office or at the construction site.

- Prequalification of potential bidders, by issuance, review and audit of subcontract questionnaire form
- Selection of subcontractors to bid on construction work
- Preparation of subcontract bid packages
- Pre-bid meeting with subcontractors, if necessary
- Receipt and opening of subcontractors' bids
- Bid evaluations (technical and commercial), production of bid tabulation summary sheet, bid conditioning for missing scope
- Negotiation with subcontractors
- Award of subcontract
- Pre-construction conference with subcontractor
- Confirmation of adequate insurance coverage prior to commencement of construction activities
- Collection of documentation related to subcontractors' bonds, letters of credit, and guarantees
- Verification and approval for payment of subcontractors' invoices, audit quantities, man-hours, markups, unit prices, hourly rates, and so on; confirmation that invoices are in accordance with the executed subcontract
- Maintenance of record log of all subcontractors' invoices
- Maintenance of shop drawing and submittal log
- Issuance of change orders to subcontractors
- Review and audit of subcontractor change orders and claims
- Maintenance of subcontractor as-built drawings
- Receipt of any required start-up manuals from subcontractors
- Release of any retention funds
- Close-out of subcontract; issue final close-out report

Contract Negotiations

Negotiating can be defined as the process of arriving at mutual compromise, or a meeting of the minds through bargaining and modifying specific elements of a contract. A negotiated agreement occurs when two parties reach a mutually acceptable compromise or agreement to a previously held position.

In the process of constructing a building or facility, negotiations must take place between all parties to arrive at a mutually satisfactory agreement for performing the work. These negotiations are usually undertaken between the following parties.

- Owner and contractor
- Contractor and subcontractor(s)
- Contractor and various suppliers or vendors
- Subcontractor(s) and sub-subcontractor(s) and various suppliers or vendors

On competitively bid construction work, such as lump sum, fixed price contracts, negotiations will generally be minimized because of all the detailed documentation available during the bid process. More extensive negotiations are usually required on cost reimbursable contracts or negotiated projects as a result of the lack of completed documentation such as drawings, specifications, and the construction contract as a form of agreement.

The usual items that need to be negotiated and resolved are cost, contract terms and conditions, insurance, warranties, coordination procedures, safety requirements, and any modifications to the contract sum. There are four basic steps in the negotiation process:

(1) Presentation of each parties' views and current position
(2) Review, evaluation, and critique of the other parties' views and position
(3) Bargaining and adjustment to each party's previously held views and position
(4) Agreement to new position

Many private industry owners carry out negotiations with several contractors before selecting one contractor to perform the work. The owner and contractor will then undertake the final negotiations and arrive at a mutually satisfactory conclusion. Similarly, after reviewing a letter of intent or an executed contract, the contractor will undertake final negotiations with the subcontractors and suppliers. The same situation occurs between subcontractors and sub-subcontractors.

The person chosen to negotiate on behalf of the owner, contractor, subcontractor, or sub-subcontractor must be knowledgeable in the construction and contracting process. This individual must be aware of all the technical and commercial aspects of the work, and should have full authority to make decisions regarding commitments that his or her organization will fulfill.

Exclusions, Exceptions, and Alternatives

On private industry construction projects, most bid document packages allow the bidding contractors and subcontractors to include exclusions, exceptions, and alternatives to their bids. When these exclusions, exceptions, and alternatives are submitted with the bid, they become conditions under which the contractor or subcontractor will perform the work. If these items are unacceptable to the owner – or in the case of subcontractors, unacceptable to the contractor – the bid may be deemed unacceptable. Normally, the owner and the contractor negotiate a satisfactory compromise solution to these exclusions and exceptions.

On many public sector construction projects, the bidding document package will state that exclusions, exceptions, and alternates will not be considered and may invalidate the bid submission.

The following is a partial list of typical exclusions or exceptions in a contractor's or subcontractor's bid:

- *Bid is based on the following drawings.* The contractor or subcontractor should list all drawings the bid is based on, indicating the drawing description, date, and revision number, or designation and date the drawing was received. In this way, any revision drawings from bid package drawings will be the basis of future change orders or the basis of a future claim.
- The bid should indicate whether *removal of any rock or underground obstructions* is included in the bid. If underground obstructions are encountered and are not indicated on any drawings, the contractor or subcontractor will need to be compensated for this occurrence.
- *Ground water* — the contractor or subcontractor should state whether the bid includes any dewatering requirements. Usually,

the contractor or subcontractor includes in his proposal a value for removing normal rain water.

- The contractor or subcontractor should indicate whether any *shift work, overtime, or weekend work* is included in the bid. If none has been allowed, the contractor should state this.
- The contractor or subcontractor should state that the bid is based on *current building and fire code requirements.* Any future changes to these regulations and codes while the project is in its construction phase could be the basis of a change order to the owner.
- Many subcontractors will exclude *scaffolding* from the bid. In many cases this would be acceptable to the contractor. The contractor and some of the other subcontractors may need to use scaffolding for different phases of the construction effort. It is sometimes better if the contractor handles and coordinates the scaffolding requirements.
- The subject of *cleanup* often causes problems or confusion on construction sites. The contractor or subcontractor should indicate what cleanup costs are included in the bid.

Private and public owners may request that alternate schemes or approaches for performing the work be included in the bid. The cost of the alternative(s) and the schedule implications are usually requested to be shown on the Bid Form or in the Attachments to Bid. Owners request alternative pricing with the hope that contractors will submit a lower bid price. The owner's reasoning is that contractors are experts in construction – they may know a better approach to constructing the building or facility since they are involved in construction every day. The owner is usually hopeful that the contractor will come up with a more optimized and less expensive construction approach. For the same reason, contractors also use this approach when soliciting bids from subcontractors.

Change Orders

A change order is a written order issued after the execution of a purchase order, contract, or subcontract. A change order can be issued between the following parties.

- Owner and contractor
- Contractor and subcontractor(s)
- Subcontractor and sub-subcontractor(s)
- Any of the above and a vendor or supplier

A change order instructs one or all of the contracting parties to increase or decrease the quantity of work; omit any part of the work; or change the quality, character, or performance of the work.

The usual procedure is for the owner's engineer or architect to issue a formal change order to the contractor. If the work to be changed has been subcontracted, the contractor will issue a change order to the applicable subcontractor(s). In the case of a purchase order, the vendor or supplier will be requested to supply the applicable cost details associated with the change order.

The best option for handling change orders is to devote a section in the owner's construction contract document package and the contractor's subcontract package to percentage overhead markup and unit price rates. For added or deleted work these rates will then be the established basis for pricing and establishing the cost value for each particular change order. If no unit rates or markups are contained in the various contracts, the parties will have to negotiate the cost of each change order, which can result in time-consuming disputes. It is usually not practical or economical to hire a different contractor to perform the change order work.

If an owner presents a large change order to a contractor, it is normal for the contractor to request additional time to perform the work. The contractor or subcontractor should inform the owner of the effect the change order will have on the completion date of the project. Numerous change orders can seriously interrupt the normal course of construction and can result in delays, budget overruns, and claims to the owner.

Bonds

In the construction industry, bonds can be used at the request of the owner as protection against any default or failure by the contractor. Bonds are typically furnished by insurance companies, which will assume the financial liability for a contractor's default or failure to perform on a construction project. An owner can invoke a bond only when there is a breach of contract by the contractor.

On private industry construction projects, bonds are used at the discretion of the owner, usually after discussions with the architect or engineer. The construction contract document package will usually indicate whether the owner requires any bonds. On public sector construction projects, bonds are required by law. In the mid-1930s, Congress enacted The Miller Act, which describes bond requirements on public sector construction work.

The cost of obtaining a bond is not usually significant enough to influence a contractor or subcontractor's bid (from 0.5% to 5% and more of bid volume). The contractor or subcontractor is most concerned with his bonding capacity or bonding ceiling. Insurance companies establish the bonding capacity of construction companies. The bonding capacity depends on the financial health and current assets and liabilities of the construction company. The contractor or subcontractor may have contracts in hand that are filling up the company's bonding capacity. The contractor or subcontractor may wish to bid on projects in which bonding capacity will be breached. The best solution is for the contractor or subcontractor to discuss and review with the insurance agent the possibility of increasing the value of the bonding capacity.

There are three basic types of bonds that are used on construction projects:

Bid Bonds

A *bid bond* is a contract between the bidding contractor, the owner, and the insurance company. The insurance company guarantees reimbursement to the owner if the contractor refuses to enter into formal agreement with the owner. The owner will be compensated for the difference between the low bid value and the eventual bid value the owner has to pay.

Payment Bonds

Payment bonds (often referred to as labor and materials payment bonds) are purchased by the contractor from an insurance company.

The insurance company guarantees payment of all labor and material costs to the contractor's subcontractors, vendors and suppliers if the contractor fails to pay these organizations.

Performance Bonds

Performance bonds are purchased by the contractor from an insurance company. The insurance company guarantees the owner payment of the costs associated with having another organization perform the construction work if the contractor fails to perform the work.

Contract Implementation

Good planning and execution are the key elements for successful contract implementation. There is no basis for control without an established strategy or implementation plan.

Figure 5.11 is an example of a contract implementation plan that an owner may use as a tool to control and monitor progress on a project. The plan indicates the phases of a competitively bid construction project. This plan may be modified for other contracting approaches.

Example of a Design and Build Agreement

The following is an example of a guaranteed maximum price (design and build) agreement or contract between an owner and contractor. The contract has an article describing any shared savings that occur on the completion of the construction work. This example can be adjusted to conform to other project applications by adding, deleting or modifying any of the articles.

Agreement for the Design and Construction of the Following Facility or Building

for

____

(Owner)

____

(Address)

Agreement: Made this day of ________ in the year of nineteen hundred and ninety ____________ .

Between __________________________ , the OWNER, and XYZ Contractors Incorporated, the CONTRACTOR.

For services rendered in connection with the following described construction project:

To complete the design and construction of ______________________________ on land owned by the OWNER, located at ______________________________ in accordance with the following contract documents.

The contract documents, which constitute the entire agreement between the OWNER and the CONTRACTOR, except for change orders issued in accordance with and after execution of this agreement, and agreed to in writing by OWNER and CONTRACTOR, are listed as follows:

__________________	__________________
Initials (Owner)	Initials (Contractor)

Contracting Implementation Plan

Project No. 4-189A
Project Name March Inc.
Revision No. 1
Date 5/4/9X

Evaluation of Owner's Needs (Months 1–5)	Bidding Phase (Weeks 1–5)	Award Phase (Weeks 1–2)	Construction Implementation Phase (Months 1–12, varies with magnitude of project)
Assessment/Synthesis	Competitive bids obtained	Recommendations	Perform total scope of work Division 1 thru 16. with a combination of direct hire and subcontractor labor
	Receive bid package from owner		
Owner Assesses needs space, headcount	Make decision to bid		Mobilization, site staff + temporary facilities
	Perform detailed quantity takeoff		Contractor performs material buy-out; Completion of project
Owner selects engineer/architect	Make decision on what to subcontract		
	Solicit bids from subcontractors/suppliers		Contractor confirms subcontractors bids
Engineer/Architect performs drawings	Perform site visit		and finalizes subcontractors
+ specification work	Hold prebid meeting with subcontractors		Contractor performs subcontract administrations, bond,
	Select most attractive bids		invoice payment, change orders, back charges, etc.
Owner/Engineer/Architect produces construction	and include in overall bid.		Perform inspection activities
contract document package.	Summarize overall estimate + review		on any required equipment; Obtain operating/start-up manuals
	with management		Review and approve
		Owner reviews contractors bids	shop drawings
Commencement of project	Review mark-ups + select profit margin	Owner negotiates with 2/3 low bidders	Perform close out activities
	Submit bid to owner		Ensure subcontractors have necessary insurance coverage.
	Qualify subcontractors	Owner awards contract	

Figure 5.11

- This entire agreement
- Preliminary design documents as listed under Exhibit A attached
- Change orders
- Written amendments to this agreement
- Final construction drawings and specifications
- Scope of work statement as listed under Exhibit B

Article 1. Definitions

1.1 OWNER is ______________________________
Address ______________________________
Authorized representative: ______________________________
____

1.2 CONTRACTOR is XYZ Contractors Inc., 123 Main Street, Anytown, New York 10500. Authorized representative:
____

1.3 CONTRACT is the Contract for the Design and Construction and is described by the Contract Documents. This Contract represents the entire agreement between the parties hereto and supersedes all prior negotiations and agreements. A modification to the Contract is a written amendment or modification to the Contract signed by both parties hereto, or a written Change Order to the Contract.

1.4 CONTRACT DOCUMENTS consist of the Contract drawings and specifications, all exhibits and amendments attached hereto, and all Change Orders issued after execution of this Contract.

1.5 CHANGE ORDER is a written modification to the Contract Documents issued by OWNER and accepted by CONTRACTOR that outlines and defines the modification to be performed.

1.6 CONTRACT PRICE means the sum specified as such in Article 7, as such sum may be increased or decreased in accordance with the terms of this Contract.

1.7 CONTRACTOR means XYZ Contractors, Inc., or the successor in interest of such company, or the assignee of such company or of any such successor.

1.8 DRAWINGS means the drawings produced by CONTRACTOR as part of the Work, as such drawings may from time to time be supplemented, amended, revised, or otherwise modified in accordance with the terms of this Contract.

1.9 SITE means the lands and other places on, under, in, at, or through which the Work is to be performed.

1.10 SUBCONTRACTOR is an individual including construction equipment and material supplier or entity who has a direct contract with the CONTRACTOR to perform any part of the Work at the Site.

______________________________ ______________________________
Initials (Owner) Initials (Contractor)

1.11 SUBSTANTIAL COMPLETION of the Work is when the OWNER and CONTRACTOR agree that construction is sufficiently complete to be occupied or utilized by the OWNER for the use it was intended to perform.

1.12 WORK comprises the completed design and construction required of the CONTRACTOR by the Contract Documents, and includes all labor, materials, equipment and other items required of the design and construction work.

Article 2. Contractor's Responsibilities

2.1 The CONTRACTOR shall arrange and supply the following services, which shall constitute the WORK. The CONTRACTOR shall supervise and direct the WORK and will be solely responsible for all construction methods, quality, and procedures.

2.2 The CONTRACTOR will provide all necessary materials and equipment, supervision, inspection, testing, labor, construction equipment, and specialty items required to execute and complete the design and construction of the WORK.

2.3 The CONTRACTOR shall comply with all applicable laws, rules, statutes and regulations of public authorities relating to the WORK.

2.4 The CONTRACTOR shall keep the site free from accumulation of waste materials or rubbish caused by the CONTRACTOR'S operations.

2.5 The CONTRACTOR shall maintain at the site one complete record set of the DRAWINGS and specifications, CHANGE ORDERS, and other modifications marked to record any changes or modifications made during construction. These shall be handed over to the OWNER upon completion of the WORK.

Article 3. Owner's Responsibilities

3.1 The OWNER shall designate a representative who shall be fully knowledgeable with the WORK, and have authority to approve changes in the scope of the WORK, render approvals and decisions promptly, and furnish information to the contractor expeditiously.

3.2 The OWNER shall furnish topographical survey maps describing physical characteristics; soils reports and subsurface investigations; legal limitations; utility locations; a property survey and the location of the project benchmark.

Article 4. Assignment

4.1 This CONTRACT shall not be assignable, and the CONTRACTOR shall not let, assign, or transfer all or any part of this CONTRACT, or any interest therein, without the prior written consent and approval of the OWNER.

______________	______________
Initials (Owner)	Initials (Contractor)

Article 5. Subcontracts

5.1 All elements of the WORK that the CONTRACTOR does not self-perform with its own labor force shall be performed by SUBCONTRACTORS. The CONTRACTOR will select qualified and competent SUBCONTRACTORS.

Article 6 Contract Time Schedule

6.1 The WORK to be performed under this agreement shall commence on or about _________________ and shall be completed within _________________ working days, on the following date _________________.

Article 7. Contract Price

7.1 The CONTRACT PRICE for CONTRACTOR'S entire performance of the WORK shall be guaranteed maximum price in the amount of $ _________________. The guaranteed maximum price shall only be increased or decreased for approved changes in the WORK.

7.2 The CONTRACTOR agrees that it shall have no claim for any amounts in excess of the guaranteed maximum price unless such Price is changed by CHANGE ORDER, or supplemental agreements, to this contract.

7.3 Immediately following completion of the WORK, the total actual cost of the completed WORK shall be compared with the guaranteed maximum price. _________________ percent of the savings shall be paid to the CONTRACTOR, and _________________ percent of the savings shall accrue to the benefit and account of the OWNER.

Aritcle 8. Payment

8.1 CONTRACTOR will issue and send an invoice for payment to the OWNER every four (4) weeks based upon a mutually agreed WORK progress to that date, less the aggregate of previous monthly payments and less ten (10) percent retainage. OWNER will make payment within thirty (30) days of receipt of invoice. Upon SUBSTANTIAL COMPLETION of the entire WORK, the OWNER shall pay a sum sufficient to increase the total payments to 95 percent of the CONTRACT PRICE.

8.2 The final payment value shall be made to the CONTRACTOR thirty (30) days after the completion of the WORK included in this CONTRACT and its acceptance by the OWNER. Before final payment is made, the CONTRACTOR shall give the OWNER evidence that the WORK is free and clear from all liens or claims.

Article 9. Indemnification

9.1 CONTRACTOR hereby assumes the sole liability and responsibility for and agrees to indemnify and save the OWNER harmless from any loss by reason of the liability imposed by law upon OWNER for damage because of bodily injuries including death, sustained by any person or persons or for

_________________	_________________
Initials (Owner)	Initials (Contractor)

any damage to property arising out of the performance and execution of the WORK under this CONTRACT.

Article 10. Insurance

10.1 The CONTRACTOR shall purchase and keep valid during the term of the CONTRACT, insurance of a form and with an insurance company satisfactory to the OWNER as follows: Workers' compensation insurance including occupational disease and employer's liability insurance covering all CONTRACTOR employees and SUBCONTRACTOR employees engaged in the performance and execution of this CONTRACT. Employer's liability insurance shall have a minimum limit of $250,000. Comprehensive automobile liability insurance covering CONTRACTOR and SUBCONTRACTOR for claims arising from owned, rented, and leased vehicles with limits of $500,000 per occurrence. Comprehensive general liability insurance including contractual and products liability with minimum combined single limits of $500,000 for bodily injury and for property damage per occurrence.

10.2 Certificates of the above insurance shall be of the occurrence type of comprehensive general liability policy and shall specify dates and expiration of coverage. A copy of these insurance certificates should be sent to the OWNER within ten (10) days before the time the work under this contract is to commence.

Article 11. Miscellaneous Provisions

11.1 The CONTRACTOR agrees that all WORK performed and all material supplied by it shall strictly comply with the laws and ordinances in force in the locality in which the WORK is located.

11.2 The CONTRACTOR agrees to be responsible for and promptly pay for all losses from thefts and other dishonest acts of its employees and its SUBCONTRACTORS and their employees.

Article 12. Warranty Period

12.1 The CONTRACTOR agrees to guarantee the WORK executed and performed under this CONTRACT for a term of one year from the completion thereof, and to make good promptly, without any expense to the OWNER, for any defects which may appear during the time of said warranty period.

12.2 The CONTRACTOR guarantees that all materials and equipment furnished under this CONTRACT will be new and in accordance with the CONTRACT Specifications, and that all WORK will be of good quality, and workmanship free from faults and defects and in strict conformance with the CONTRACT DOCUMENTS.

Article 13. Arbitrations

13.1 In case the CONTRACTOR and OWNER fail to agree to matters under this CONTRACT, then the matter, if mutually agreed to by OWNER and CONTRACTOR, shall be referred to a board

____________________ Initials (Owner)

____________________ Initials (Contractor)

of arbitration. The decision of this board shall be final and binding on both parties to this contract.

Article 14. Governmental Requirements

14.1 If any requirements of the CONTRACT shall contravene or be invalid under the laws of the state or jurisdiction, such contravention or invalidity shall not invalidate the whole CONTRACT.

14.2 Throughout the execution and performance of this CONTRACT, the CONTRACTOR agrees not to discriminate against any employee or applicant for employment because of race, color, religion, sex, handicap, or national origin.

14.3 This CONTRACT is to be interpreted in accordance with the laws and statutes of the State of New York.

Article 15. Termination

15.1 If at any time after execution of this CONTRACT, there shall be filed by or against CONTRACTOR in any court, a petition for bankruptcy or insolvency or for reorganization or for the appointment of a receiver of all or a portion of CONTRACTOR'S property or if CONTRACTOR makes an assignment for the benefit of creditors, this CONTRACT, at the option of OWNER, may be cancelled and terminated immediately by written notice to CONTRACTOR. OWNER may retain as liquidated damages, the funds of any outstanding CONTRACTOR'S monthly progress payments.

15.2 OWNER may, at its own option, cancel this CONTRACT at any time on thirty (30) days written notice to the CONTRACTOR for any reason. Should OWNER cancel this CONTRACT, it is agreed that CONTRACTOR will be reimbursed by OWNER for all costs of the WORK, expenses, and obligations incurred by CONTRACTOR in connection with this project prior to the date of such termination notice.

15.3 In the event of termination by OWNER, OWNER shall accept CONTRACTOR'S commitments under all incomplete purchase orders issued, and outstanding at the time of receipt of said termination notice by CONTRACTOR.

15.4 If the OWNER fails to make payment when due to the CONTRACTOR, the CONTRACTOR may give written notice of the CONTRACTOR'S intention to terminate this CONTRACT and recover from the OWNER payment for WORK and CHANGE ORDER work executed and for losses sustained upon materials, equipment, and construction equipment, including reasonable profit, legal fees and any applicable damages the CONTRACTOR has been subject to.

______________________	______________________
Initials (Owner)	Initials (Contractor)

In witness thereof, the parties have hereunder set their hands and seals as the day and year first written above.

Owner: ______________________________

By: ______________________________

Title: ______________________________

Attest: ______________________________

Title: ______________________________

XYZ Contractors, Inc.

By: ______________________________

Title: ______________________________

Attest: ______________________________

Title: ______________________________

Exhibit A: Preliminary Design Documents

Drawings:

Number & Revision		Title	Dated
A–1	Rev "B"	Floor Plan & Elevations	10/25/9X
A–2	Rev "B"	Sections & Details	10/25/9X
M–2	Rev "B"	HVAC Plan	10/25/9X
P–3	Rev "C"	Plumbing Details	10/25/9X
E–4	Rev "B"	Electrical Details	10/25/9X

Specifications:

Document No. A-0001 Rev "B" dated 10/24/9X, consisting of specifications outline contained on a ten-page document. (Divisions 1 through 16.)

Exhibit B: Scope of Work Statement

The CONTRACTOR shall provide all necessary ______________________________

____

____

____

____

*Note: Additional articles can be incorporated into this CONTRACT as the situation may dictate. However additional articles must be approved by both parties to this CONTRACT.

______________________________ Initials (Owner)

______________________________ Initials (Contractor)

Standard Forms of Agreement

There are a number of preprinted standard forms of agreement for construction services that are used throughout the U.S. The forms that are widely used within the construction industry are those created by the American Institute of Architects (AIA) and the Engineers Joint Contract Documents Committee (EJCDC). The EJCDC document is jointly produced by the National Society of Professional Engineers, the Consulting Engineers Council, the American Society of Civil Engineers, and the Construction Specifications Institute. Another form of agreement is produced by the Associated General Contractors of America (AGC).

In general, all the preprinted forms of agreement are well prepared, based on many years of use and application. There are certain advantages to using these forms. They have been widely accepted within the construction industry; many times they have been subjected to legal reviews and interpretation in many court cases. The forms are generally well known and understood throughout the construction industry, and are for the most part general in nature.

No one standard form of agreement can address all the specific needs and requirements of every construction project. Usually, these forms are selectively modified or added to by the owner and contractor to reflect the actual location, site conditions, and owner requirements that will prevail throughout the construction process.

Many large and mid-sized private industry owners have, over the years, developed their own forms of contract for construction work. These forms are typically hybrids of the three previously discussed forms of contract. Many owners have added specific language and clauses into their forms of contract to reflect particular operating practices and procedures that are encountered at their facilities or operating plants.

Owners who do not possess their own standard form of agreement might hire a lawyer to draft a form of agreement for the project. This approach, although not widely used, does have certain benefits and advantages to the owner.

All parties engaged in the construction process – owners, contractors, and subcontractors – are well advised to obtain legal advice on any forms of agreement they are considering being party to. The vast majority of forms of contract can be negotiated and modified to better serve the needs and requirements of contracting parties.

The following is a partial list of preprinted contract documents that are used extensively in the construction industry.

AIA Documents

- A101 Owner-Contractor Agreement Form – stipulated sum (4/87)
- A101/CM Owner-Contractor Agreement Form – stipulated sum, construction management edition (6/80)
- A111 Owner-Contractor Agreement Form – cost plus fee (4/87)
- A201 General Conditions of Contract for Construction (4/87)
- A305 Contractor's Qualifications Statement (12/86)

EJCDC Documents

- 1910-8-A-1 Standard Form of Agreement between Owner and Contractor on the basis of a stipulated price
- 1910-8 Standard General Conditions of the Construction Contract

The American Institute of Architects (AIA) General Conditions of Contract for Construction Form A201 is widely used throughout the construction industry. The form basically acts as a operating/coordination/administration procedure between the two contracting parties. It describes the operating procedure under which the construction work will be performed and administrated. It also describes the coordination role the architect performs on the project. The following is a listing of the fourteen articles contained within this form.

Article 1 General Provisions

Article 2 Owner (definition) (information required of Owner)

Article 3 Contractor (definition) (obligation of Contractor)

Article 4 Administration of the Contract

Article 5 Subcontractor

Article 6 Construction by Owner or by Separate Contractors

Article 7 Changes in the Work

Article 8 Time

Article 9 Payment and Completion

Article 10 Protection of Persons and Property

Article 11 Insurance and Bonds

Article 12 Uncovering and Correction of Work

Article 13 Miscellaneous Provisions

Article 14 Termination and Suspension of the Contract

It would be prudent for contractors and owners to purchase a copy of this document and familiarize themselves with its contents. It is available from the American Institute of Architects, 1735 New York Avenue, N.W., Washington, DC 20006. It is also available at many architectural bookstores.

Challenges in Contracting

The particular needs of the owner usually dictate which contracting method will be appropriate in the construction of a facility or building. The owner will consider many factors, many of which have been discussed throughout this chapter, when making a choice regarding contracts.

Demands of the Lump Sum, Fixed Price Contract

The most demanding contract a contractor faces is the lump sum, fixed price contract. The main elements of this challenge are as follows:

- The bid due date. Usually the contractor has only four to six weeks to complete a detailed cost evaluation of the proposed building or facility.
- If he or she is successful in submitting a bid to the owner, the contractor must commit substantial financial and labor resources to the project and mobilize a team to commence the construction work, usually immediately.

- The contractor must manage and orchestrate his or her resources and subcontractors' resources to complete the total construction effort at a cost below his submitted lump sum, fixed price contract value, and within the project time schedule, at the same time complying with the quality requirements of the owner.

Successful performance of these tasks requires careful and detailed planning. To grow and remain profitable in today's economic environment, contractors need to develop project priorities and apply today's advanced management techniques to their construction projects.

Important Considerations for Working with Contracts

The following items relate to contracting methods and techniques that can serve the interests of both owners and contractors.

- Decisions regarding which type of construction contract to use are influenced by completion date requirements, the completeness of the detailed design, and the owner's need to change or modify the work in progress.
- Carefully planned and prepared forms of agreement, conditioned to the specific scope and location attributes of the facility or building, result in positive results that are reflected in the bid submissions and bid values.
- The articles or clauses contained within the body of the contract that usually cause problems or disputes on construction projects are the additional work or changes in the work, payment, and completion date. These elements as well as the definition of the scope of work tend to be responsible for the majority of problems or disputes. It is advantageous to both owner and contractor to carefully review and understand these particular articles.
- Unit prices for additional work and % markup for overhead and profit should be requested by the owner and should be part of the construction contract document package.

All procedures and methods described in this chapter may be modified and adapted to meet the specific needs of owners, contractors, and subcontractors who are engaged in construction contracting.

Chapter 6

International Purchasing, Contracting, and Subcontracting

Chapter 6

International Purchasing, Contracting, and Subcontracting

As recently as one or two decades ago, many construction-related firms in the U.S. had little or no interest in working overseas because of the perceived complexities and problems involved. It was easier to work and remain in this country. As a result of many political and economic events – such as the creation of a single European market, the dramatic changes in the former Soviet Union and Eastern Europe, the reunification of Germany, the dramatic world population growth, the end of the arms race, the rapid growth of certain Southeast Asia economies – the 1990s and beyond promise to be a time of opportunity for working overseas. Many construction firms have pursued overseas work and have found that it was well worth the effort.

Construction organizations can expand their revenue and sales base by performing work overseas; in many cases, organizations that have ventured overseas have found that international business represents a substantial part of their total sales revenue and profit. International work also gives a construction organization a certain stature in the marketplace; it implies that the company is capable of undertaking a certain complex level of work.

The decision to undertake work in a foreign country should not be taken lightly. The acceptable level of financial risk and current and future investments should be carefully evaluated. Realistic assessments must be made regarding the dexterity, capabilities, and strengths and weaknesses of the company. Particular considerations for work in a foreign country include the geographic area to be penetrated, business opportunities, types of products or services that may or may not be available, competition, and profit goals.

The global marketplace for engineering, design, construction, and construction materials and equipment is a high risk, fiercely competitive arena, in many ways similar to the current situation in the U.S. domestic market. However, for U.S. organizations that successfully penetrate and perform in this arena there are real opportunities for substantial professional and financial growth. The following is a list of some of the benefits of working in the international construction arena.

- Overall sales volume of the organization is increased.
- Potential new markets and opportunities for the organization's services and expertise are opened up.
- General growth and expansion plans are augmented.
- Fixed overhead costs are distributed over greater sales base and volume.
- Downturns in the U.S. domestic marketplace could be counterbalanced.
- Various U.S. methods, technology, systems, and management techniques and practices are used and exploited to magnify advantages in underdeveloped countries.
- Job opportunities and growth potential for existing and new staff are generated.
- Projects for existing and new U.S. clients who require work overseas can be cultivated.
- The risks associated with being in one specific marketplace are minimized by the involvement in diverse markets.
- Overall return on investment is enhanced; profit return is optimized.
- A contribution is made to the correction of the U.S. trade deficit imbalance.

This chapter assumes that a corporate evaluation and decision to pursue international work has been made. The information presented applies to any organization in the U.S. or other industrialized nation performing construction work abroad. Such an organization might be a manufacturing/owner-type company; an architectural or engineering firm; a construction management firm; a managing contractor/general contractor or specialist subcontractor; or a joint venture enterprise, including host country organizations, government agencies, or any combination of the above.

This chapter also presumes that the U.S. organization is undertaking construction work on a medium- to large-sized construction project in a remote area of a developing country. These types of projects might include chemical plants, steel mills, mines, military complexes, oil and gas fields, pipelines, and housing complexes. Any of these facilities could require roads, hospitals, schools, housing, jetties, airports, and so on.

The primary purpose of this chapter is to provide information regarding the purchase of materials, equipment, and construction services for international projects. We explain and elaborate on the decisions to be made, the knowledge and insight required to make these critical decisions, and the sources of the information necessary to perform this undertaking.

Historic and Future Trends

Many world events have influenced the growth of opportunities in the global marketplace. Rapidly expanding markets, along with the dramatic population growth currently taking place, will continue to fuel opportunities for construction work overseas. Listed below are a number of the world's geographical areas including former trading or military alliances and their estimated populations for the year 1995. These population estimates illustrate the magnitude of the global marketplace. Listed in parentheses with each area or continent are some of its major countries by population.

Geographical Area	1995 Estimate of Population in Millions
U.S.	260
Canada and Mexico	110
Europe (Germany, U.K., France, Italy, Spain, etc.)	650
Former Eastern Block Countries (Poland, Hungary, Romania, Bulgaria, etc.)	150
Russia	320
Asia (China, India, Indonesia, Pakistan, Bangladesh, etc.)	3,200
Africa (Algeria, Egypt, Kenya, Nigeria, Zaire, etc.)	700
South America (Brazil, Argentina, Peru, Chile, Venezuela, etc.)	380
Australia/New Zealand/Pacific Islands	30
Estimated World Population in 1995	5.8 billion

The United Nations predicts that the world population will grow from 5.8 billion in 1995 to over 7 billion in the year 2010. This reflects more than a 20% increase in a 15-year period. The world has never experienced this kind of population growth. For example, in these 15 years Brazil's population is estimated to increase from 145 million to 205 million; India's population is expected to increase by 50% from 800 million to 1.2 billion; and Nigeria's population will more than double from 105 million to 215 million. The U.S. population will grow by more than 20 million from 260 million in 1995 to over 280 million by the year 2010. For these and all the other nations of the world there will be a great demand for construction work and services, including new roads, bridges, schools, housing, hospitals, airports, food production plants, and manufacturing facilities to support and sustain the rapid and unprecedented population explosion.

The expenditure for new construction work in the U.S. in 1993 is estimated to be $310 billion. This value does not include engineering, design, owner construction management costs, or contractor construction management services. This value covers all of the five main types of construction work: residential, commercial, civil engineering (e.g., roads, bridges, and dams), industrial, and institutional. Although it is difficult to quantify the total worldwide construction market, certain researchers in this field have calculated that it could be in the range of five to ten times the value of the U.S. market. This could mean a staggering yearly construction expenditure in excess of $2,000 billion. Hence the growing opportunities for construction organizations to perform work all over the world.

Planning for International Construction Purchasing: A Checklist

The discussion in Chapter 2 (The Overall Procurement Plan) becomes especially meaningful and relevant on overseas construction projects, especially those located in developing countries. The additional coordination requirements resulting from climatic conditions, isolated geographical locations, language, culture, work week differentials, construction methods and practices, and religious practices can all significantly influence project plans and goals. Early planning is a must for overseas projects.

The following list may be used as a checklist that can be added to and modified for specific needs. These questions and their answers will assist senior management and the purchasing professional in strategizing and directing the procurement planning process. Answers to some of these questions can typically be obtained from the targeted or host country's U.S. embassy. This list is not all-encompassing but will assist in the development of a strategy.

- In what form must purchase orders, contracts/subcontracts, and consulting agreements be in? Should they be verbal or written? In what language (local or some other)? Can two languages be used, or is a certified translation needed? Must they be passed before a notary public or the host country's equivalent? Must they be published in local or national newspapers and/or displayed in government offices or facilities?
- To what degree are the parties to a contract free in their choice of law and option of deliberation and subsequent decision? Can the legal venue for the settlement of a dispute be decided in a country other than where the work is being performed?
- At what stage does title or transfer of materials and equipment pass on a sale?
- Is there in existence clarification and decisions on contracts? Is there a system of laws, decrees, and ordinances or statutes?
- To what degree is arbitration accepted in the targeted country? Are the decisions of an arbitrator panel upheld?
- Can contracts be modified or adjusted by consent and agreement between the contracting parties? What is the usual procedure?
- Does the host country recognize the following forms of business entities?
  - Corporations
  - Public limited companies
  - Private (or limited-liability) companies
  - Partnerships
  - Joint ventures
  - Consortiums
- Do these forms of business have a successful track record? Are the foreign country's government departments and private companies familiar with these types of business organizations?
- Are there any restraints on these forms of organizations? Can foreign nationals operate and control these forms of organizations? Is any special permission required?
- Does any form of business have any distinct advantage such as nonpayment or deferment of taxes, subsidies, start-up grants, training assistance, or new business incentive programs?
- Does the host country impose its own import and export regulations or does the country belong to a union or trade body such as the European Economic Community (EEC) or the European Free Trade Association (EFTA)?
- Does the host country extend "most favored nation" status to the U.S.? What is the U.S. status on this matter?
- Are any import licenses necessary for construction materials, major equipment, or related products? If so, what are the regulations governing the issuance of such import permits and licenses?

- Are there any overall restrictions or stipulations on value, quantity, or type of imports?
- What is the composition of the target country's import duty program? What appraisal method is used to value products being imported?
- Are export permits and licenses required from the U.S. or other third-party country for importing items into the host country? If so, how are they acquired? What is the usual amount of time needed to obtain these permits and licenses?
- Does the host country provide financing for new facilities and plants through grants, low-interest loans, incentives, tax deferments, or exemptions?
- Does the host country allow repatriation of capital and profits back to the U.S. or another country, or must capital and profit remain in the host country for subsequent reinvestment? Is foreign exchange subject to any form of regulations or government control?
- Does the host country offer preferential or special terms or conditions? If so, how are these terms and conditions obtained?
- What recourse or protection does a U.S. organization have against expropriation or nationalization of its business assets?
- What are the various tax scales and rates and how are they calculated and applied to:
  - Locally hired personnel?
  - Expatriate employees?
  - Consultants?
  - Profits?
  - Losses?
- To what degree are the working methods and conditions monitored by the host government?
- What are the host country's regulations on minimum wages or salaries; maximum number of hours worked per week or day; paid vacations and statutory public holidays; overtime and shift differentials; discharges or terminations?
- What are the host country's current views and policies toward privatization or nationalization of specific industries?
- Does the host country have any regulations in effect that limit the percentage of business ownership by foreign nationals?
- How is depreciation of property, equipment, and other assets handled?
- Can a U.S. organization own property or assets in the host country?
- What are the requirements for visas, work permits, and entry and exit procedures?
- What is the current value added tax (VAT) rate? (VAT is a form of sales tax added to the cost of consumer products; the rate varies from country to country.) What other sales taxes are in effect?

When first considering overseas work, it is important to realize the distinctions in the classifications of countries. Generally, Second and Third World countries can be classified as the countries other than the 24 members of the Organization for Economic Cooperation and Development (OECD). The 24 member countries consist of the 19 most developed Western European countries along with Canada, the United States, Japan, Australia, and New Zealand. These countries are usually referred to as First World countries or the industrialized or developed countries.

Surveying the Target Overseas Country

The project planning and procurement efforts will be greatly strengthened by the performance of early surveys within the target country. This survey as it relates to materials and equipment should be carried out by senior members of the procurement group. The survey is a crucial step in the preparation of a comprehensive international procurement buy-out plan, especially if the U.S. organization has not had previous or recent experience in the host country.

The survey of the targeted country should cover the following major subjects:

- Availability of local labor, contractors, subcontractors, consultants, engineers, and architects
- Possible local joint venture partners
- Wage rates and salaries by craft or profession
- Names of local vendors, suppliers, and contractors, and areas of expertise
- Delivery times and costs of major pieces of equipment and all other bulk material items that are manufactured in the host country
- Contact names and addresses for local/national banks
- Contact names and addresses for shipping facilities, airports, storage areas, transport companies
- Office, housing, and hotel facilities; cost and rental information regarding these facilities
- Work permit requirements, resident permits, visas, exit permits, and so on
- Local employment agencies, if they exist
- Government agencies, addresses, telephone numbers, contacts
- Local attorneys or solicitors
- Tax structure of country and names of local accounting firms knowledgeable on international tax matters
- U.S. embassy address and contacts
- Any requirements regarding a stipulated percentage of locally manufactured materials and equipment that must be installed and used on the project (usually monitored very closely by the government officials)
- Regulations concerning whether materials and equipment from other countries can be used on the construction project
- Telecommunication systems, telephone, telex, fax machines; installation time frames and costs
- Computer usage and availability
- Import/export and customs requirements
- Local permits, approvals, licensing, and regulations concerning purchase of automobiles
- Insurance requirements (personal, medical, and property)
- Medical facilities, doctors, emergency treatment
- Medical facilities open to expatriates
- Regulations on export of profits back to U.S.

- Safety requirements
- Inspection services
- Currency exchange rates and current and projected escalation rates
- Local taxes, levies, import duties, tariffs, business and occupation taxes
- Business practices; work and installation practices; productivity comparisons between target country and U.S. labor rates (this information will be very helpful when estimating the total project cost)
- Regulations concerning whether expatriate workers can bring families to country
- Where children of expatriate workers may attend school
- Cost of housing and rental property rates
- Requirements of temporary work camps
- Cost of freight and transportation
- Airline, train, and public transportation costs
- Typical costs of food, clothing, and related services

It is not unusual to encounter major delays in certain developing countries because harbors and transport systems are either too small or antiquated. These facilities need to be examined if materials and equipment are to be brought into the country in this manner.

The survey effort usually requires a team or task force approach. The team should generally consist of senior members of the procurement, estimating, planning, engineering, and project management groups. Survey plans require a step-by-step approach and should be analyzed in detail before commencement of the survey task. Detailed questionnaires should be developed, and travel plans and appointments should be made in advance and confirmed before any meetings.

The information that is obtained from the list above should augment the information obtained from the host country's U.S.-based embassy. This combination of information will greatly assist the procurement planning effort. The survey can provide a database for possible future work in a particular geographical location. This data can also be made available to any joint venture partners, specialist subcontractors, or additional project partners or team members.

Financing an International Construction Project

Once the decision to undertake overseas work has been made, the next logical consideration is financing. The first step in obtaining financing is for the construction firm to approach its local commercial bank. Over 200 U.S. banks have international banking departments that employ specialists who are familiar with specific countries and the various types of transactions and financing arrangements that are prevalent within those countries. These larger banks are typically located in major U.S. cities, and are generally knowledgeable about export requirements, finance matters, and currency exchange transactions. If the construction organization's bank does not have an international department, one or more of the following organizations can usually provide advice or information:

- Eximbank Export-Import Bank
- Small Business Administration
- U.S. & FCS (U.S. and Foreign Commercial Service) district office

- World Bank
- Overseas Private Investment Corp.
- Agency for International Development

These organizations can provide information regarding possible attractive financing sources as well as valuable analysis and research data. Most of this information is free. The functions of most of these organizations are to promote and assist U.S. companies performing or considering overseas work.

If the contractor can dictate the payment terms on the overseas construction project, he or she will have a distinct advantage. The optimal way to finance a project is to front-load the payment terms; for example, the contractor might receive 25% of his or her bid value when the contract is signed, and additional payments could be linked to the achievement of certain key milestones or main elements of work. The contractor should endeavor to finance the project with the owner's funds in this way.

On the other hand, the construction organization may have sufficient financial strength to establish a commercial line of credit from its own banking sources, or the organization may have adequate funds to finance the project.

Favorable payment terms by the owner may in certain circumstances make a particular project more attractive than it would be otherwise. This will become apparent to the contractor as he or she reviews the owner's bid package and the stipulated payment terms.

Payment periods for overseas projects can vary in duration from 30 to 120 days from the date of the contractor's invoice. The financing costs associated with longer payment periods can be considerable. The length of time that financing is required increases costs in proportion to the duration of the project. Interest rates and bank fees change constantly, and the contractor should include these financing costs in the bid submitted to the owner.

Considering Foreign Currencies and Exchange Rates

Local currency must be used for local purchases such as labor, materials, equipment, and services. The local funds must be available at the required dates; otherwise, payments to contractors, subcontractors, vendors, and suppliers can be delayed. The use of an imprest or overdraft bank account that contains a constant fixed amount of funds, established jointly by the owner, the government agency, and the contractor can be a factor in avoiding this type of situation. (An imprest account is a loan or advance payment fund, usually financed by the owner.)

Public Works Projects: The usual method of financing overseas public works projects (e.g., roads, bridges, dams), especially in Second and Third World countries, is to use funds provided by one of the United Nations-type lending institutions. Some of these institutions include Inter-American Development Bank (IDB), International Bank for Reconstruction and Development (IBRD), and Crown Agency (CA). There is usually some kind of delay during the life of one of these loans. These lending institutions are funded by various countries and organizations. Delays are typically caused by red tape, bureaucratic error, and late payments from countries supplying part of the funding appropriation. These delays can cause serious cash flow problems for the project.

Private Projects: When the overseas construction project is for private industry (i.e., financed by a Fortune 500 or 1000 privately owned company), many of the funding delay problems described earlier do not occur. The contractor's main concern is obtaining enough local currency on a regular basis to pay local contractors, subcontractors, vendors, and suppliers. Because of bank and government regulations in some countries, the time required to exchange U.S. dollars or another Western currency to local funds can take anywhere from 3 to 14 days. This problem can be overcome by planning for the various transfer and banking routines that are required for exchanges, and monitoring the cash flow requirements on the project.

For all types of projects, the contractor must ensure that payments due to him or her from the owner or government agency are received at the milestones or dates stipulated in the contract. Otherwise, a domino effect can occur, causing payment delays to vendors, suppliers, and subcontractors, and eventual claims and disputes.

Establishing a Bank Account Overseas: For a contractor to open a bank account in a foreign country, government permission must usually be obtained and all the necessary forms must be prepared and approved. Particularly in some Second and Third World countries, the establishment of a bank account by a foreign organization can be very time-consuming. If the construction project's expenditures are made in a number of different currencies it is vital that exchange rates and conversion records are maintained for all currency exchange transactions. This information will be the basis of any currency impact statements, which indicate variances in the pre-established exchange rates stipulated in the contractor's original estimate or bid. This allows the contractor to claim back from the owner any significant currency movements or major fluctuations that may occur during the execution of the project.

Researching Cultures, Languages, and Ethics

Business methods, work practices and ethics, religious views and beliefs, dietary customs, humor, family values, and acceptable dress can vary widely from one country to another, especially in some developing nations. A contractor should strive to research the area and/or question people who have been to the country to gain a basic understanding of the business environment and culture, management views and attitudes, business practices, and consumer preferences.

Avoiding Conflicts

Although the contractor need not "go local" when performing business in a foreign country, he or she should be aware and open to local customs and business practices. The following advice should be followed where possible; failure to do so may cause offense or lead to possible conflict.

- Discussions about current or former political leadership or the political system of the host country should be avoided.
- Religious opinions or beliefs usually should not be discussed or debated.
- Preaching the merits and prowess of the U.S. or other Western

countries can cause friction with certain individuals and should be avoided.

- Hard sell negotiating techniques used in the U.S. and Western Europe are not appropriate in certain Second and Third World countries; these methods can cause offense.
- Discussions or debates on the qualities and merits of various political systems should be avoided.
- Refusing certain local foods or alcoholic beverages may be excused if the refusal is given in a tactful and friendly manner. Visitors to overseas countries should adopt the eating and drinking habits of the locals whenever possible.
- In certain parts of the world, engaging in small talk is expected prior to commencing detailed business discussions. This approach should be followed where possible; allow five to ten minutes of small talk.

Dealing With Language Differences

Generally, U.S. organizations performing work in non-English-speaking countries are required to conduct day-to-day business in two languages – English and the host country's language. The requirement of dealing in both English and the host country's language dramatically increases the scope, volume, and complexity of work for the procurement management group. The costs associated with translations, explanations, and clarifications for English-speaking personnel can be significant, however, contractors should invest the time and money in teaching their English-speaking staff the particular foreign language that is needed for an overseas project. This makes good sense both with current and future clients and with the local staff.

Many large U.S. Fortune 500-type companies that construct facilities overseas insist that all contractual and business activities are conducted in English. This of course makes the execution of the construction project much easier for the U.S. organization and more difficult for the local construction community.

When the construction project is performed using the host country's language, all project and related procurement matters (correspondence, request for proposal, drawings, contracts, specifications, purchase orders, invoices, etc.) must be written and expressed in that language. All key documents must also be clearly understood by the contractor's project team. There are certain ways of overcoming this problem. The best method is to have all project forms and procedures compiled in both English and the local language. This enables the expatriate and local staff to effectively perform their particular procurement functions.

The various language disparities that can occur overseas can, of course, add to final costs of the project. Additional staff, decreased productivity, and duplication of certain activities can all influence the cost and duration of the project. These variables must be calculated and included in the bid or budgeted cost of performing the work.

One of the best ways of overcoming the language barrier is to hire local staff or expatriate procurement staff that are bilingual or at least have a knowledge of the particular language. The more individuals on the project who understand the two languages, the better the communication will be on the project.

Considering Ethics

The topic of business standards and ethics includes the morals and honesty of each person associated with the construction project, as well as the standards of behavior of the local business community. Such standards should be fully explored and understood before commencing any business activities. Certain foreign countries have substantially different business practices from the standard operating principles of the U.S. For example, commissions, finder's fees, and award fees are normally used in many overseas countries, but are not usually used in the U.S. or OECD countries. The U.S. contractor must strive to develop operating procedures and business practices that safeguard the interests of both the U.S. organization and the owner of the construction project. At the same time, the U.S. contractor must be able to adjust and adapt to the business practices, standards, and ethics of the host country.

Before undertaking any business or procurement activities in a particular country, it would be prudent to research the host country's business methods and practices and the moral and ethical manner in which business is conducted. U.S. organizations should be aware of U.S. government rules and regulations such as the Foreign Corrupt Practices Act legislation (FCPA); anti-diversion, anti-boycott, and anti-trust laws; and Food and Drug Administration (FDA) and Environmental Protection Agency (EPA) restrictions. These regulations outline conduct guidelines for U.S. organizations and citizens performing business activities in foreign countries. A publication entitled *Basic Guide to Exporting*, published by the U.S. Department of Commerce, is an excellent reference source on this particular subject.

Establishing Procurement Operations in a Foreign Country

Construction projects located in more remote developing foreign countries typically establish the project office and material storage facilities at the project site or as close as possible to the site. Sometimes, however, the material and equipment procurement office is located near the capital city or main provincial city or town, where the various government agencies or departments that monitor importation of materials and equipment usually conduct business. It helps greatly if the procurement office is located close to a major seaport and/or international airport. The appropriate location helps to optimize logistics related to receiving equipment and materials.

Characteristics of Office and Storage Areas

The construction project's offices and the temporary warehouses, laydown areas, storage sheds, and material staging areas are usually required only for the duration of the project (i.e., 12 to 30 months) and are therefore not up to the standards or characteristics that would be expected for a permanent business office. If the U.S. organization is contemplating having an ongoing presence in the overseas country, it would be logical to rent or purchase a permanent office as well as the necessary storage and staging facilities from which future business operations could be conducted.

Special Considerations for Working in Developing Countries

Services such as potable water, electricity, gas, sewers, and roads are sometimes taken for granted in the U.S. and Western Europe. In Second and Third World countries these services may be either nonexistent, extremely poor in quality, or limited in service output. For example, local telephone lines available for use can be insufficient or not available. Telephone service typically is subject to overloads and breakdowns, and in many cases the operators speak little or no English. In underdeveloped Second and Third World countries obtaining telephone, telex, or fax systems can be a major undertaking, taking anywhere from two to six months to obtain. In some countries, fax machines are not used at all. Portable C.B. phones may be the short-term solution to this problem.

Lack of electricity may necessitate the use of portable electric generators. Weather extremes – temperature, rainfall or lack of rainfall, snow, and so on – can also have an effect on the construction effort. All of these considerations need to be evaluated and planned for.

Hiring Personnel for an International Project

Selecting and hiring the right people with the necessary skills and experience are essential to the success of any construction project. Individuals with previous overseas procurement experience will prove to be invaluable.

The organizational charts that are depicted in Chapter 3 (Purchasing) can be used or adapted for use in overseas procurement activities. The job descriptions can also be adapted to meet host country and construction project requirements.

Local Staff

All U.S. organizations considering overseas construction work should understand the host country's labor ordinances, rules, and laws before recruiting local workers. In hiring local personnel, the following points should be considered.

- Certain foreign labor laws prevent employers from assigning employees to work beyond a confined narrow category of work. As a result, a task may require more employees than the U.S. contractor might expect.
- The vast majority of labor laws in many overseas countries make it almost impossible to terminate or dismiss an employee except for the most extreme cause.
- The labor laws in many overseas countries are biased toward the employee. This situation tends to encourage an attitude of detachment or aloofness in the employee, resulting in overstaffing, waste, absenteeism, and inefficiency.
- Productivity in some countries is extremely low. A task that takes one hour to perform in the U.S. may take three or four hours to complete.

Employees in certain liberalized European countries have vacations in excess of seven weeks per year. French workers, for example, are entitled to a minimum of five weeks vacation when they start a new job, plus 11 national and religious holidays. Some Italian companies allow newlyweds an extra two weeks vacation above and beyond their normal vacation allowance. In some overseas countries, businesses close for a two-week period over Christmas. The situation in Japan and some of the Far East countries is just the opposite. Most workers are entitled to three weeks or more vacation a year, but many take less because of the value placed on hard work and diligence to one's job and employer.

In some countries, it is not unusual to see employees leaving work at 3:00 in the afternoon. Standard work hours vary significantly from one country to the next. For example, in South Korea the average work week is 54 hours; in Japan the workers toil an average of 47 hours per week; in the U.S. the average work week is 40 hours; and in some European countries the work week is less than 35 hours long.

In many cases standard work weeks are mandated by the local government. Time off for maternity leave, illness, and public holidays in many countries is much longer than the typical time allowed in U.S. organizations. Unemployment benefits in Europe are also very generous. Typically they decrease over time, but do not cease regardless of how long an individual is out of work. All of these situations influence the productivity of workers.

Expatriate Staff

The practice of using existing U.S.-based procurement employees, either based in the U.S. or temporarily relocated overseas, or hiring experienced expatriates on overseas projects is highly recommended. Experienced professionals can have a positive influence on the overall success of the construction project, particularly in the procurement and buy-out effort. On many overseas projects expatriate workers are typically from North America, Western Europe, Australia, or New Zealand. Over the last five years, expatriate workers from the Philippines, South Korea, Egypt, and Pakistan have been working in some of the major Middle Eastern countries. These individual can also be considered expatriates.

The following points should be considered when reviewing the need for U.S. or other country expatriates on an overseas construction project. Although hiring expatriates is beneficial to the project, the process can be rather complex.

- Obtaining the required work permits, visas, and authorization for expatriates can be an extremely long process, sometimes taking as long as six months. Certain overseas countries try to discourage the use of expatriates; because some of these countries constantly experience high unemployment rates, they prefer to use local workers. Some overseas countries allow a limited number of expatriate workers on a particular project because of this situation.
- The costs of keeping an expatriate in a foreign country can be as much as 60% to 120% more than the costs of employing workers in the U.S. cost of living adjustments (COLA's), per diem rates, tax differentials, and tax equalization rates must be established prior to the commencement of any overseas assignment. U.S. nationals living and working overseas are subject to U.S. taxes plus local country taxes.
- Procurement activities are typically on the project schedule's critical path on an overseas construction project, with the potential problems related to obtaining work permits, visas, authorization etc. It could be beneficial to temporarily assign U.S. home office purchasing personnel on visitors visas (if possible) to the project to get the procurement effort off the ground. In this way any

critical long lead equipment items can be ordered and fabrication work can begin.

- Individuals assigned to overseas projects should be informed about the various problems and conditions that they are likely to encounter: there may be only local-language television, radio, newspapers, and magazines; some countries do not allow alcoholic beverages; Western-type food may be unobtainable; Sundays may be workdays; and 60-hour weeks may be the norm. They should be given periods of rest and relaxation, extended vacation periods, and salary raises and/or completion bonuses to compensate for the inconveniences and difficulties they are sure to experience.
- Expatriate personnel who are transferred to construction projects in developing countries should have temperaments that will allow them to perform effectively in the overseas country. They should be willing and able to train the locally hired procurement staff where necessary.

Organizing the Overseas Purchasing Plan

The crucial first step in planning and executing the purchasing effort for an overseas construction project is to develop a consensus among key project participants. These key participants are usually the owner or government agency, lending institution, engineering firm, contractor, and various project managers, team members, construction engineers, and procurement managers.

The purchasing group should be established at the international construction site location as early as possible. Equipment and material deliveries to the project site are usually on the critical path of the project schedule, so it is vitally important that the purchasing group be in place and working early in the project. This is especially important if the construction project has a fast track schedule or if a bonus or penalty clause is part of the overall contract.

Decisions must be made reasonably quickly concerning which materials and equipment will be imported and which will be manufactured and purchased locally. An approved list of vendors and suppliers must be prepared, specifications and standards for the project must be determined, and dates for the completion of the facility or building must be established. All of these decisions plus many other determinations need to be evaluated and finalized.

When all of the pertinent issues have been resolved, a preliminary purchasing plan can be formulated. Figure 6.1 is a responsibility matrix chart for procurement activities; it is relevant to the compilation of the preliminary purchasing plan. This chart can be used as a management planning tool. The objective of the matrix chart is to specify which party within the project team is responsible for specific purchasing activities. This is a Level 1 (summary level) responsibility matrix chart for procurement activities related to a large overseas construction project. More specific details could be shown in Level 2 and Level 3 charts.

Figure 6.2 shows the typical procurement cycle for purchase orders. This flow chart can be used for U.S.-based projects or international construction projects.

The methods discussed in Chapter 3 can be adapted for purchasing activities on international construction projects. Standardized

Responsibility Matrix for Project Procurement Activities

Project No. 7-034 X
Project Name Wrights Inc.
Revision No. 1
Date 7/10/9X

Construction Project Equipment/Materials	Owner-Furnished Equipment		Managing Contractor-Furnished Items		Admin. Bldg. Contractor-Furnished Items		Pipeline Contractor-Furnished Items		Pump Station Contractor-Furnished Items		Railroad Contractor-Furnished Items		Comments
	H.O. Purch'd	Field Purch'd	H.O. Purch'd	Field Purch'd	H.O. Purch'd	Field Purch'd	H.O. Purch'd	Field Purch'd	H.O. Purch'd	Field Purch'd	H.O. Purch'd	Field Purch'd	
Dryers (3)	X												Managing contractor to inspect and expedite (18 months delivery). Owner provides dryers and grinders.
Grinders (3)	X												
Towers (4)			X										
Vessels (4)			X										
Compressors (2)			X										
Pumps (50)			X										
Heat Exchangers (10)			X										
Storage Tanks (12)			X										
Concrete (A)				X		X (A)		X (A)		X (A)		X (A)	(A) Managing contractor will set up concrete batch plant, will supply other contractors, and will backcharge contractors.
Steel Piping				X		X		X		X		X	
Main Pipeline Pipe							X						
Building Materials													
• Lumber				X		X		X		X		X	
• Paving/Stone				X		X		X		X		X	
• Blockwork/Brick				X		X		X		X		X	
Elevators (2)					X								
Structural Steel				X (B)	X (C)								(B) Main structural steel to process plant + piperacks (C) Main structural steel to admin. building
Pipe (less than 12")				X		X		X		X			
Valves			X			X		X		X			
Fittings				X		X		X		X			
Supports				X		X		X		X			
Main Elec. Equipt.			X (D)										(D) Managing contractor will purchase all electrical equipment and materials and will free issue to other contractors
Instrumentation			X (E)										
Insulation				X		X		X		X			
Paint				X		X		X		X		X	
Temp. Main Camp			X										(E) Instrumentation same as (D).
Temp. Pipeline Camps							X						
Temp. Pump/Stat. Camp									X				
Laboratory Equipt.	X												
Railroad Equipt.											X		
Office Equipt.		X											
Temp. Office Equipt.				X		X		X		X		X	
Plant Truck		X											
Plant Autos		X											
Spare Parts		X											
Plant Start-up Materials				X									

Figure 6.1

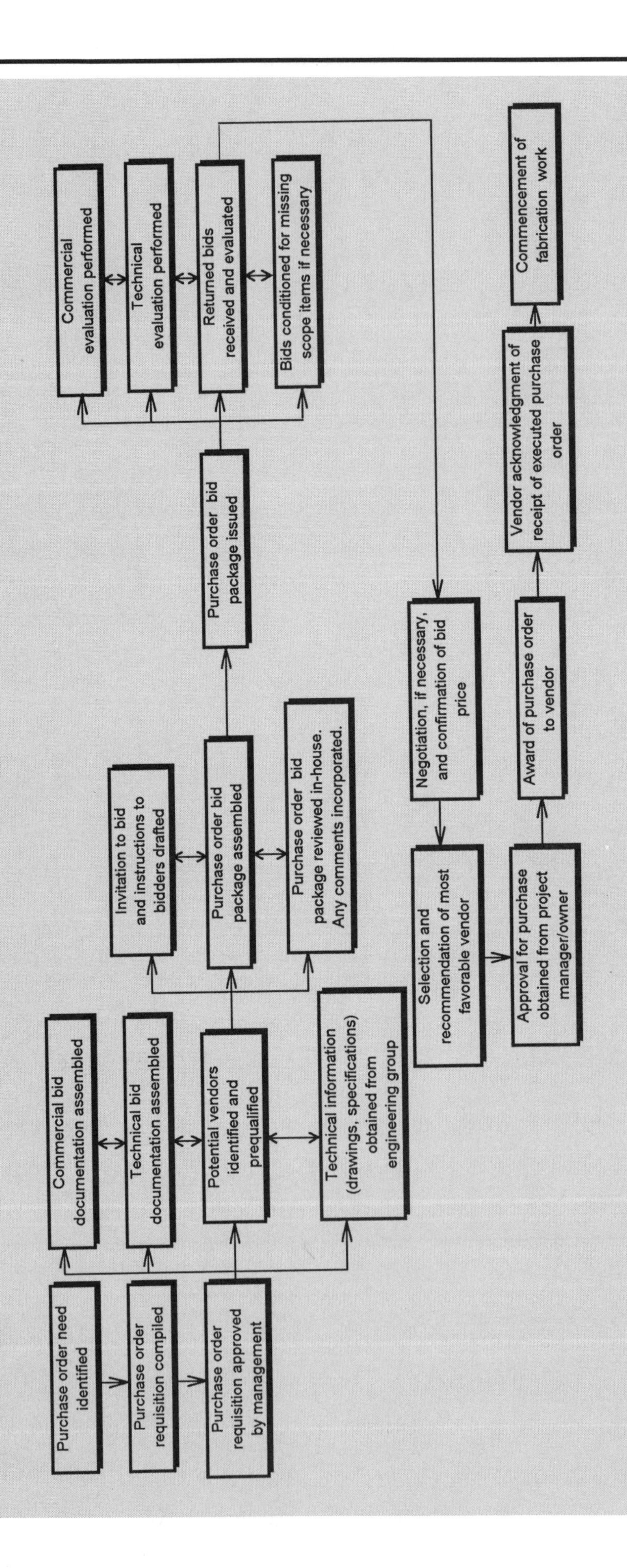
Purchase order need identified
Purchase order requisition compiled
Purchase order requisition approved by management
Commercial bid documentation assembled
Technical bid documentation assembled
Potential vendors identified and prequalified
Technical information (drawings, specifications) obtained from engineering group
Invitation to bid and instructions to bidders drafted
Purchase order bid package assembled
Purchase order bid package reviewed in-house. Any comments incorporated.
Purchase order bid package issued
Commercial evaluation performed
Technical evaluation performed
Returned bids received and evaluated
Bids conditioned for missing scope items if necessary
Negotiation, if necessary, and confirmation of bid price
Selection and recommendation of most favorable vendor
Approval for purchase obtained from project manager/owner
Award of purchase order to vendor
Vendor acknowledgment of receipt of executed purchase order
Commencement of fabrication work
The Purchase Order Procurement Cycle

Figure 6.2

purchasing forms are available in Appendix A; these forms can also be modified and used for certain international projects.

Deciding Where to Purchase Materials and Equipment

On many international construction projects in which a U.S. contractor or owner is performing the procurement effort, it is sometimes beneficial to purchase some key long lead delivery items from the home office. This situation usually depends on the size and magnitude of the purchasing effort and how long it takes for delivery of equipment and materials to the site. On large and long duration construction projects, it is usually more practical and efficient for the U.S. organization to set up a procurement office at or close to the construction site.

It is not unusual for the host country's government to demand that a large percentage of equipment and materials for the construction project be purchased from local vendors, suppliers, and manufacturers. Some host countries stipulate that at least 70% of all equipment and materials for a project be bought within the country, which can make the buy out process very difficult and complicated. For example, during the late 70s and early 80s the Canadian government stipulated that a high percentage of all equipment, materials, and other services being used for large Canadian oil and gas construction projects had to be obtained from Canadian companies. The government agency that implemented this procedure would monitor, review, and approve all bid tabulations, purchase orders, and contracts.

If the host country requires that local industry be used, the purchasing group may be involved in significant and detailed reviews, meetings, and negotiations with government agencies in addition to their usual purchasing duties.

Acquiring Materials and Equipment

Before procurement activities can begin on an overseas construction project, a list of local vendors and suppliers should be researched, analyzed, and formulated. The requirements of the host country's government and of the various organizations involved in the project should be fully understood. The quality of the finished products produced by local vendors and suppliers must be assessed. Criteria for approval of materials and equipment must be established, and milestone events related to these activities need to be determined and included in the master schedule.

Creating Vendors and Suppliers Lists

When researching and making contacts with potential vendors or suppliers it would be prudent to evaluate the operational, technical, and business capability of these potential vendors and suppliers by performing on-site investigations. Certain vendors, suppliers, and contractors, particularly in Second and Third World countries, may have difficulties understanding or producing work to the quality and standards that are the norm for industrialized nations and in compliance with specifications, drawings, or other engineering data. Engineering and production capabilities can vary from adequate to poor. In many cases, vendors and suppliers in Second and Third World countries have not been exposed to stringent specifications and inspection requirements that are typical in developed countries; their concepts of manufacturing, testing, and quality control may be insufficient. The procurement management team needs to recognize and plan for this possibility. Many Second and Third World countries who have some industrial manufacturing capability still do not have established directories, listings, or catalogs for local vendors or suppliers. Researching and establishing lists in some of these foreign countries can be a difficult and time-consuming undertaking. The following organizations can usually assist in creating a vendors and suppliers list.

- Embassy/consulate commercial departments
- Local joint venture partners
- Professional architectural/engineering firms
- Government agencies and departments
- Professional societies
- Union organizations
- Equipment and materials brokers
- Trade groups
- Manufacturing councils
- Publishers of directories with trade information including telephone numbers/contact persons and business addresses

Working With Materials and Equipment Manufactured in Second and Third World Countries

One of the most demanding and intricate problems encountered by purchasing groups when working in Second and Third World countries is the stipulation by the host country that locally manufactured materials, services, and equipment must be used on the construction project. Listed below are some of the specific problems in using host country construction materials, services, and equipment.

- The host country's suppliers, vendors, and manufacturers are often protected by import regulations, duties, taxes, and tariffs. Many of these organizations have a habit of substantially increasing their selling price because they know that there is limited opportunity to maneuver by the U.S. (or other First World) contractor.
- When the host country has only a small manufacturing base, there may not be much competition among local suppliers, vendors, and manufacturers. Therefore, project cost and schedule goals could be jeopardized.
- Local suppliers, vendors, and manufacturers may find it difficult or impossible to supply materials and equipment that conform to the quality standards required in the project specification.
- Local suppliers, vendors, and manufacturers are usually set up to supply the needs of a limited or finite market. If a construction project requires a great amount of materials and equipment, it may be too much for the local manufacturing base to handle at one time. Delays, quality problems, and cost increases can be the end result.
- Local suppliers, vendors, and manufacturers usually give first preference on manufacture and delivery of materials and equipment to local customers and clients. They are not prepared to imperil business with regular customers.
- Local suppliers, vendors, and manufacturers may not be able to supply materials and equipment in the quantities and at the production rate required for the construction project. The project completion date may need to slip in response to this situation.

The events mentioned above can also occur when performing work in some of the 24 industrialized countries but generally this is a very rare occasion.

Working With Imported Materials and Equipment

Materials and equipment for the overseas construction project can be purchased by either the locally established purchasing group or the U.S. organization's home office purchasing group. Typically, the local purchasing group acquires bulk materials and commodities (concrete, structural steel, pipe, cable, and building finishes) and less sophisticated equipment that are manufactured in the host country.

Complicated or sophisticated equipment (compressors, distillation towers, instruments, elevators, and computerized control systems) is usually manufactured in one of the 24 OECD countries. This equipment could be purchased by either the local purchasing group or the U.S. home office, but it is often better for the home office purchasing group to handle the purchase. Sophisticated equipment items are typically long lead delivery items that are easier to purchase from an established office, rather than a local group that may be experiencing start-up and coordination problems in the early stages of the project.

Local regulations regarding the import of materials and equipment should be fully researched and understood. These requirements include (but are not limited to) import licenses, customs procedures, tariffs, duties, marine insurance, shipping regulations, bills of lading, and certificate of origin documentation. Delays in receiving imported materials and equipment at the job site will have a serious effect on the construction project's completion schedule and cost goals. The U.S. Department of Commerce issues detailed information regarding imports to and exports from foreign countries. Spare part requirements for sophisticated equipment should also be evaluated. These spare parts (e.g., pump impellers, turbine blades, special machine parts and components) should be ordered as soon as possible and delivered with the equipment.

Expediting for an International Project

For any overseas construction project, an expediting group must be established and detailed expediting procedures must be formulated early in the project's life cycle. The duties of the expediting group are to monitor the progress of vendors, suppliers, manufacturers, and contractors and to report on any delays or potential problems that will jeopardize the delivery of materials and equipment. The number of expediters required for a particular project will depend on the size, scope, and schedule of the project as well as the magnitude and level of expediting required and the degree of material and equipment sophistication.

The expediting function for an overseas construction project is very similar to expediting for U.S.-based projects. The main difference is that vendors, suppliers, manufacturers, and contractors in Second and Third World countries are typically less concerned with achieving strict schedule and delivery dates than their U.S. or OECD counterparts. This cultural difference can have serious ramifications for the construction project. To overcome this potential problem, a detailed and comprehensive expediting implementation program should be developed for host country materials, supplies, and equipment.

The expediting program requires extensive contact with vendors and suppliers – surveillance and monitoring, daily or weekly reports on actual physical progress, visits to vendor or fabrication shops, and telephone follow-up calls. An efficient expediting program will ensure that materials and equipment are delivered to the job site on the contracted delivery dates with minimal delays to the schedule.

Expediting contact, surveillance, and monitoring activities for projects in OECD countries are usually not as comprehensive or detailed as those required for projects in Second and Third World countries. Expediting for construction projects in developed countries can be performed by either the U.S. home office, a European branch office, a qualified consulting firm, or an independent expediting consultant.

International Quality Assurance and Quality Control Programs

A major element of the international procurement effort that needs to be planned for and constantly monitored is the need to purchase materials and equipment that conform to the requirements of the project drawings and specifications. The U.S. organization's project management, purchasing, and engineering groups, with input from the owner, must determine which industry codes and standards will be enforced on the construction project.

Materials and equipment must arrive at the construction site in good functioning condition. The materials and equipment also have to be installed according to specified tolerances and the correct installation procedures. The means of ensuring that these two important objectives are realized is achieved through the use of quality assurance and quality control programs.

It is especially important that high standards of materials and equipment are specified for international construction projects; this will alleviate some of the possible problems discussed earlier. The QA/QC of installation work is achieved when qualified construction management personnel perform the necessary inspection, audits, testing, and monitoring activities on the work being installed by the various contractors and subcontractors at the job site.

Quality Considerations in Second and Third World Countries

Construction projects in Second and Third World countries can suffer severe complications from failures of key items or components of materials and equipment. Key items or components may take weeks or months to obtain from OECD countries; typically these countries supply the more sophisticated materials and equipment for international projects. Obtaining specific equipment or a certain component in a short time can be extremely expensive and disruptive to the project. In its most severe form this problem can cripple or close down a construction project. This situation can be minimized by having a large inventory of spare parts at the construction site; this solution can also be expensive, however.

Local workers or suppliers are usually not knowledgeable about installation services or techniques used in industrialized countries. Projects in undeveloped countries usually require a higher level of inspection and surveillance of both the fabrication methods used by the local organization and the delivery and installation sequences.

Inspection/Surveillance Teams

Quality assurance and quality control are often difficult to control or achieve when local suppliers, vendors, manufacturers, and contractors are inexperienced or have never worked on construction projects where industry standards and codes such as American Society for Testing and Materials (ASTM), Construction Specifications Institute (CSI), American Society of Mechanical Engineers (ASME), British Standards (BS), and National Electrical Manufacturers Association (NEMA) are used.

When local vendors, suppliers, and contractors do not have adequate QA/QC procedures, an inspection/surveillance team may be formed by the U.S. contractor and key individuals associated with the local contractor or fabricator. This team helps the local organization conform to the quality and delivery objectives of the construction project. The inspection/surveillance team coordinates with the other project team members (e.g., project management, procurement, and engineering groups, and local vendors, suppliers, and contractors) to ascertain whether project specifications and requirements can be waived or modified to allow a more expedient fabrication and delivery cycle. Relaxing requirements can often achieve a more cost-effective approach by the local organization, allowing the project to be completed within the pre-established milestones and cost budget.

Inspection of Imported Materials and Equipment

Equipment and materials that have been shipped or air transported from overseas countries to the project job site should be inspected upon arrival. The bill of lading or packing list should also be checked and audited. This inspection will uncover any damage or loss that may have taken place during the transportation phase. If damage has occurred during transit, damage reports should be compiled immediately. Damage reports should describe the damage in detail, including photos or videotapes of the damage, dates, times, names of witnesses, and other pertinent information.

Coordinating Transportation of Materials and Equipment

The transportation of construction materials and equipment to an overseas construction project is a procurement activity with many variables. Different countries, different documentation and inspection requirements, transit across third-country borders, and many other factors must be considered and planned for. The movement of equipment and materials from the permanent place of manufacture or fabrication to the overseas job site requires logistical support from the procurement group and other project groups. Other parties who need to be involved include freight forwarders, transport companies, local government departments, local police officers, and local highway departments.

Traffic and transportation activities must be incorporated into the construction project's execution plan. The delivery of equipment and materials in accordance with the project schedule is vital to the overall success of any construction project, and it is naturally a more complex process in overseas projects. Consideration should be given to transportation, packing and unpacking, loading and unloading, shipping, weather delays, export and import procedures, documentation, and all of the required inspection records. Figure 6.3

is a typical traffic and expediting flow chart that details the various steps that need to be considered when transporting equipment and materials to an overseas location.

The transportation group is responsible for the safe and timely delivery of equipment and materials from the point of manufacture or supply to the project construction site. On small or mid-sized international construction projects this function might be performed by the project manager, the planner, or, sometimes, the expediter. This group should use the most efficient and cost-effective form of transportation available (e.g., air, truck, rail car, ocean container, cargo ship, or barge). The transportation group should maintain communication with all other project groups, advising them of the estimated time of arrival (ETA) and of any improvements or delays in deliveries; in this way the construction

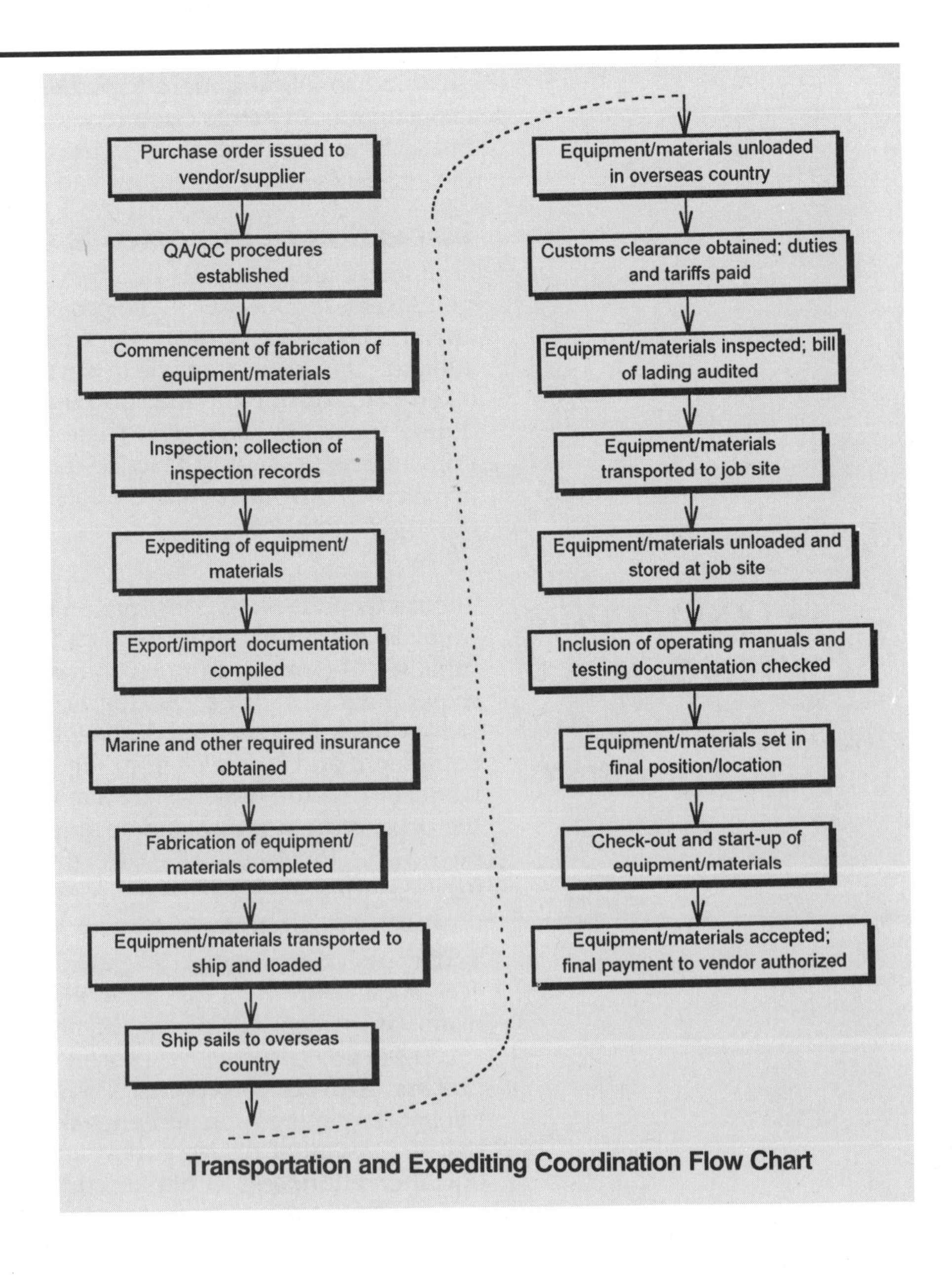

Transportation and Expediting Coordination Flow Chart

Figure 6.3

group will be in a better position to schedule and execute their work and prepare for any delays or problems that are likely to occur.

In some situations the vendor or supplier may be responsible for the delivery of materials and equipment to the construction site. In this case, the transportation group would have only a monitoring role to play.

Transportation Within the Foreign Country

Transportation of materials and equipment within Second and Third World countries can be an arduous and time-consuming undertaking. Some of these countries have the capability to manufacture and fabricate certain less sophisticated equipment and material items. However, overall transportation systems, including railways and road systems, may have serious size and weight limitations. Some equipment and materials could be too heavy or too large for the country's railways and roads; these items may have to be broken down into smaller, more manageable sizes (i.e., modules, skid-mounted units, or pre-assemblies).

Railway and transportation companies are often nationalized or state-controlled government organizations that are usually inefficient and inadequately equipped and maintained. Many of the roads and bridges in some less developed foreign countries are in a constant state of disrepair or are too narrow and winding for the movement of large or mid-sized transportation loads. Often, paved roads do not exist; dirt roads are the only means of transportation. Contractors may have to build temporary roads to gain access to the project sites in very remote areas. For example, when the large oil pipeline project was being constructed at Prudhoe Bay in Alaska, hundreds of miles of temporary stone and gravel roads, as well as bridges and river crossings had to be constructed, maintained, and eventually removed when the project was completed.

River or coastal transportation by barge or ship may be a possible solution to railway and road transportation problems. If this approach is considered, contact should be established with local barge operators and river transport companies to ascertain availability, weight limitations, cost, insurance requirements, and so on.

Functions of the Transportation Group: When the overseas construction project is located near a major city or seaport, and adequate road or railway transportation is available, the transportation group may be required only on a spot or part-time basis. However, if the project is extensive and located in a remote region, the transportation group will need to plan, coordinate, and monitor some or all of these necessary functions:

- Harbor operations, including wharfage space bookings, docking, off-loading operations, and temporary storage
- Rental of warehousing and jetties; payment of harbor fees
- Rental of barges, tugs, ground transportation
- Staging areas, rigging, craneage, temporary equipment
- Rental of trucks, vans, trailers, and storage sheds
- Transportation of the construction labor force to and from the site each day

- Transportation of food and related consumable items to remote temporary camps, offices, fabrication facilities, and staging areas
- Air transportation operations, helicopters, cargo planes and all necessary support operations associated with the items above

The transportation group's management and supervisory personnel must be chosen with great care. They must be experienced and knowledgeable in overseas work, especially in Second and Third World countries. Supervisors must be capable of training and educating the locally hired staff in all facets of transportation coordination and support. They must be familiar with all forms of management reporting; import, export, and transportation documentation; safety requirements; crane operations; traffic control; dispatching; warehousing; rigging; maintenance; and equipment operations.

Import of Materials and Equipment to the Foreign Country

There are many factors to consider when importing materials and equipment to an overseas country. Many foreign countries maintain a barrier of importation requirements that must be overcome; these regulations can require significant time, effort, and a great amount of documentation and associated procedures. This situation is very likely to cause logistical delays.

The barriers are often in place to protect the country's manufacturers, vendors, and suppliers, and to assist in the control of imports into the country. Such regulations serve to promote, protect, and assist local industries. This situation will never change; the answer is to work around the problem in the most effective and responsive manner. In addition to the foreign country's regulations, government export requirements must be conformed to. Problems and delays will occur if these procedures and requirements are not satisfied. Guidance can be obtained from many sources, including the overseas government agency responsible for imports and the U.S. Department of Commerce, which produces many fine publications on the subject.

In general, the U.S. organization performing work overseas must pay close attention to all the documentation pertaining to export and import requirements. All necessary paperwork must be fully understood and compiled in the correct manner. Failure to conform to all requirements will cause delays, problems, and potential increases in costs. Some overseas import officials are very aggressive in looking for errors in forms and other documentation. In this way they hope to discourage the importation of materials and equipment. A freight forwarding organization can often assist in dealing with all the documentation related to export and import issues.

Warehousing

When performing construction work overseas, it is essential that plans be developed for site warehouse facilities, including personnel procedures, material and equipment storage requirements, inventory control, bonded warehousing requirements, and any just-in-time materials programs. If warehousing facilities are not available or required, materials and equipment should be stored in a safe location that allows construction workers easy access but still permits efficient inventory control. It is important that materials and equipment be stored close to the final installation location, with suitable protection from theft and the elements.

Contracting, Subcontracting, and Administration for Overseas Projects

International construction projects are subject to the same demands and challenges encountered on U.S. projects. Potential problems include bad weather, material shortages, design errors, lack of correct materials and equipment, accidents, labor shortages, economic factors, delays and schedule restraints. These problems can be magnified many times over when a U.S. contractor, design firm, or owner is performing construction work overseas. The building or facility must be designed in accordance with the host country's latest standards, regulations, and building codes. In addition, every construction project, whether domestic or foreign, is unique and requires a steep learning curve at the start of the project. Every project is designed to fulfill the owner's particular needs.

The construction process for international projects is subject to a variety of internal and external forces. For example, the construction team can include a combination of government agencies, owners, engineers, architects, general contractors, and subcontractors. These individuals or teams are all new to one another and will always differ from one project to the next. Unlike a manufacturing process in which certain products are mass produced every day, the learning curve in construction is never optimized, especially on overseas projects. The many construction team scenarios, together with language and cultural differences, are hurdles that must be quickly and successfully overcome. This is especially the case in international contracting, subcontracting, and administration.

International Construction Contracts

The usual procedure for international construction work is for a general contractor or engineering, design, procurement and construction management (EDPCM) firm to enter into a contract with a government agency or owner. The contract will usually describe in detail the scope of work to be performed as well as each party's specific obligations. The method by which the contractor is chosen, the type of contract used, and the contractual requirements of the contractor can vary greatly.

Standard Forms of Contract: In the U.S., many commercial construction projects are executed under the terms and conditions of the American Institute of Architects (AIA) form of contract. The AIA form is a tried and trusted method that U.S. commercial builders, owners, and architects are well versed in. In Western Europe, similar forms of contract exist. In the United Kingdom, commercial construction projects are often performed according to the Joint Contracts Tribunal (JCT) form of agreement; civil engineering projects (e.g., roads, bridges, tunnels) use the Institute of Civil Engineers (ICE) conditions of contract.

Standard contract forms are used throughout industrialized Europe and by the majority of the 24 members of the OECD. They are often based on the methods and requirements of each country's national bodies of architects or civil engineers. These forms are similar in many respects to the AIA and JCT forms of contract. However, in many Second and Third World countries such standard forms do not exist, so the contractor may have to establish some form of contract appropriate for the particular project.

Contracts Established by Owners: Many industrialized construction projects (manufacturing plants, factories, chemical plants, refineries, steel mills, and so on) in the U.S. and overseas are performed according to a contract established by an owner or, in some cases, a government agency. While this is not usually a problem, it is strongly recommended that any U.S. contractor, design firm, or owner must *fully understand* whatever type of contract is being used. Lawyers or solicitors familiar with both construction work and local laws and regulations should be hired to review any contract for an overseas construction project.

Chapter 5 (Contracting, Subcontracting, and Administration) covers the importance of contracts and the different types of contracts in detail. These contracts may be adapted for overseas applications.

Subcontracts

All subcontracts, whether international or domestic, should be very similar in both content and style to the form of contract between the contractor and the owner. The subcontract brings two parties together for the performance of a given scope of work in accordance with a set of contract documents, specifications, and drawings. Although the owner or government agency usually has no direct contractual arrangement or communication with the subcontractor, it is good practice for the contractor to use the owner's contract as the basis for the subcontract. Passing the same project requirements on to the subcontractor is a much more satisfactory arrangement than using the standard preprinted subcontract forms that are prevalent throughout the construction industry both at home and abroad. A more detailed discussion about subcontract agreements is included in Chapter 5. A typical subcontract procurement flow chart is depicted in Figure 6.4. This sample can be used for both international and domestic U.S. projects.

Contract Administration

Details of contract administration are discussed in Chapter 5. The comments and examples contained in Chapter 5 (bidding, qualifying subcontractors, negotiations, change orders, bonds, and so on) are applicable to both international and domestic U.S. construction projects.

Deciding to Pursue Overseas Work

Many of the special situations and potential obstacles in international construction projects are created by specific cultural practices, customs, and business practices. Not all of the problems discussed in this chapter apply to every international construction project, but every contractor should be aware of the potential for setbacks caused by language barriers, business and cultural differences, and import and export requirements. Understanding and overcoming these factors will substantially enhance the success of the procurement process for any international construction project.

The procurement group should be prepared to deal with different methods of procurement, unusual terms and conditions, and different

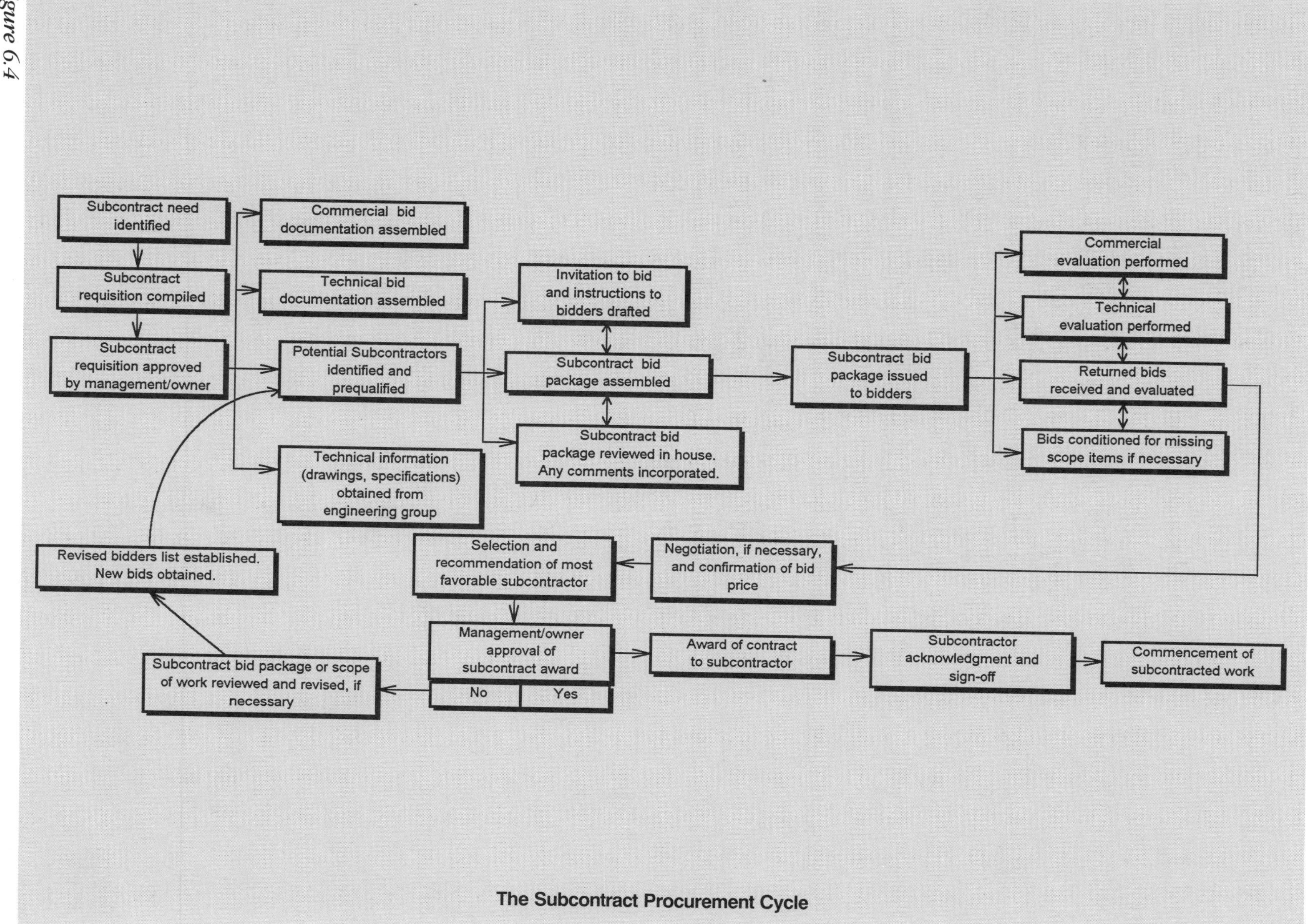

Subcontract need identified
Subcontract requisition compiled
Subcontract requisition approved by management/owner
Commercial bid documentation assembled
Technical bid documentation assembled
Potential Subcontractors identified and prequalified
Technical information (drawings, specifications) obtained from engineering group
Invitation to bid and instructions to bidders drafted
Subcontract bid package assembled
Subcontract bid package reviewed in house. Any comments incorporated.
Subcontract bid package issued to bidders
Commercial evaluation performed
Technical evaluation performed
Returned bids received and evaluated
Bids conditioned for missing scope items if necessary
Negotiation, if necessary, and confirmation of bid price
Selection and recommendation of most favorable subcontractor
Revised bidders list established. New bids obtained.
Management/owner approval of subcontract award
No
Yes
Subcontract bid package or scope of work reviewed and revised, if necessary
Award of contract to subcontractor
Subcontractor acknowledgment and sign-off
Commencement of subcontracted work
The Subcontract Procurement Cycle

Figure 6.4

business practices. Organizations considering overseas work must also remember that out of the approximately 180 countries that are presently represented in the United Nations only three use the old English Imperial measurement system (feet, pounds, cubic yards, and so on). These three countries are Burma, Liberia, and the United States. The other 177 countries use the System International (S.I.) metric system of measurement. The Metric Conversion Act of 1975, as amended by the Omnibus Trade and Competition Act of 1988, establishes the modern S.I. metric system as the preferred method of measurement in the U.S. Former President Bush signed an Executive Order on July 25, 1991, mandating that all new and updated federal publications and drawings make the transition to metric units by September 30, 1992. Today, many federal government construction projects are being designed, purchased, and constructed using the metric system. It is anticipated that the U.S. will be fully converted to the S.I. system by the end of this century.

The international construction and procurement effort is more demanding and entails greater coordination and potential risk than domestic U.S. construction projects. This chapter does not intend to imply an image of doom or foreboding. Quite the opposite is the case. The potential rewards and growth opportunities for organizations contemplating or currently performing work overseas are considerable; many are described in the preceding pages.

Globalization begins with redefining one's basic thinking; the first order of business for decision makers in both private and public industry is to think beyond national borders to embrace the concept that all of the world's markets comprise one global marketplace. There may be no better time than now to pursue international construction work.

Chapter 7

Selling Construction Products and Services

Chapter 7

Selling Construction Products and Services

The previous six chapters of this book have discussed in detail the practices and procedures of purchasing construction materials, equipment, and services. This chapter departs from the purchasing arena and focuses on the selling of construction products and services. Although the processes of construction purchasing are significant, it is also important to recognize that selling or obtaining sales of construction products and services is a very important activity. In many ways the two functions are closely connected. This chapter will shed some light on the issues and processes related to selling construction products and services.

Salesmanship is one of the most vital and dynamic forces in the free market world today. Salesmanship is the necessary spark of industrial operations and production; it is the fuel of commerce and business. Almost all manufacturers and suppliers have some form of sales force. The typical manufacturing organization would quickly be out of business without some channel or sales forum for its products.

All over the world, there is a great demand for more and better educated and motivated sales personnel. In every free market country, business organizations strive to obtain the most capable and competent sales representatives. This demand will continue to grow as the world's population and markets for products expand in the years ahead.

Over the last 10 to 20 years, manufacturers and product suppliers have come to realize the importance of marketing and selling their products and services using the latest techniques. The professional salesperson, held in check by tradition and ethics from directly marketing and selling his or her own specific services and skills, will typically get business in an indirect manner: by joining certain clubs or societies, by creating a wide circle of friends and business contacts, and in large measure by relying on the testimony and goodwill of past clients and contacts.

It is salesmanship that keeps the factories and manufacturing plants in operation, filling the trucks, railcars, ships, and airplanes and dispatching them from city to city, from coast to coast, and across the oceans of the world. The salesperson is the conduit connecting the generator (demand) with the business motor (supply). He or she is the person who dispatches articles from the point of production to be used by the consumer. The salesperson's work effort can mean the difference between success or failure for any business enterprise.

It is fair to say that the wheels of industry and of all business would come to a complete halt without the efforts of the many professional and dedicated sales representatives selling and obtaining sales of their products, services, and goods in the global marketplace.

Construction Selling Defined

There are many inaccurate definitions of the term *salesmanship*. Many people believe that salesmanship is the marketing and selling of a product to someone who does not really need it. This statement is far from true. Real salesmanship firmly represents the merits of the product being sold and at the same time promotes future goodwill related to the product. Salesmanship can be defined as the art of applying knowledge about a product and selling that product at a profit. The assumption is that the buyer will use the product to produce a profit or other desired result.

Salesmanship, as it relates to the construction and engineering industries, is the art and science of selling products or services that depend, either partly or totally, on a technical and engineering description or specification in their selection, application, use, and installation.

The total process of the manufacture and production of construction-related products relies on sales orders from end users of the product. The efforts of the designers, manufacturers, and assemblers are put together and must eventually be sold at a profit. If this procedure does not take place in a satisfactory manner, loss of profit will occur and the business may eventually fail. In the final analysis, the salesperson is the designer, manufacturer, and producer of sales orders.

The Successful Salesperson

The ability of the salesperson to create sales supports the production and manufacturing infrastructure of the organization he or she represents. To be successful, the salesperson must have a reasonable technical understanding of the product he or she is selling. He or she must also be aware of the economic facets of the marketplace.

To be successful in selling construction products, the salesperson must have a total grounding in the basics of selling. He or she must have the ability to demonstrate to the buyer that the product or service satisfactorily meets or surpasses its intended purpose or application.

The salesperson must use technical knowledge, experience, and skill to reach a satisfactory economic conclusion. In doing so he or she must apply not only technical considerations but also business acumen and knowledge. The work of a sales representative combines the knowledge and skills of an engineer, designer, manufacturer, user, and business owner. In addition to these, a successful sales representative must have the ability to explain, demonstrate, persuade, and convince the potential buyer of the merits of the product.

The efforts of a salesperson can be considered pragmatic and visionary. The salesperson's work is in many ways aimed at devising methods of producing products and services in better, more efficient ways. The salesperson educates industry and the marketplace on new and better ways of producing and eventually selling new products and services. He or she looks at the problem of selling construction products and services with a combination of a technical/engineering point of view and a business/profit point of view, reaching a course of action only after carefully considering both facets. These two points of view should be focused on finding the optimal way to furnish the required product or service to the eventual end user or buyer.

Construction Products and Services to be Sold

To have a clear understanding of how salesmanship relates to the huge construction and engineering sector of worldwide commerce and industry, we must visualize the vast range of construction products, goods, equipment, and services that we see around us every day. Listed below is just a small sampling.

- Construction equipment (e.g., cranes, bulldozers, trucks, portable generators)
- Civil engineering products (e.g., paving, stone, underground pipe)
- Concrete products (e.g., cement, lumber, reinforcing bars)
- Masonry products (e.g., bricks, blocks, mortar)
- Metal products (e.g., structural steel, metal decking, bolts)
- Wood and plastic products (e.g., lumber, fasteners, plastics)
- Moisture-thermal control products (e.g., roofing, insulation, siding)
- Doors, windows, and glass products, door and window furniture
- Finishes (e.g., paint, tiles, wall treatment products)
- Conveying system products (e.g., elevators, escalators, hoists)
- Mechanical products (e.g., pipe fittings, specialist equipment products)
- Electrical products (e.g., cable, conduits, switches)
- Furnishings and specialities (e.g., furniture, drapes, blinds)
- Engineering and design equipment and materials (e.g., pumps, compressors, heat exchangers, distillation towers)

Each of these products is unique and has its own segment or niche in the construction marketplace. No one salesperson can become an expert on every product or service. Many sales representatives specialize in one specific segment of the industry: selling windows or selling structural steel products, for example. Thus the work of a salesperson in the construction industry is typically focused on one or a few related products or services. In this way, the salesperson can better serve the specific needs of the manufacturer and the buyer.

The Importance of Knowledge

A successful fisherman must know the location of the fish as well as their habits and feeding behavior. Similarly, the sales representative's efforts are futile unless they are aimed toward current and future selling opportunities. Locating potential buyers is often one of the most demanding and difficult tasks in the sales field. Failure to identify current and future potential buyers is the problem most frequently encountered in sales work.

Segments of the Construction Industry

As mentioned earlier in this book, the construction industry is divided into five main segments or categories of construction. These segments are:

- Residential construction: single and multi-family homes, repair and renovation construction, upgrades, remodeling, and reutilization of existing facilities or buildings
- Commercial construction: office buildings, retail establishments, hotels
- Heavy civil construction: highways, bridges, tunnels, jetties, dams, sewerage treatment plants
- Institutional construction: government facilities, schools and universities, health care facilities, prisons
- Industrial construction: chemical plants, refineries, manufacturing facilities, factories

A particular product or group of similar products has either a vertical or horizontal demand in each construction segment. For example, a specialty distillation tower can be sold only to petroleum refining or chemical companies or to engineering firms doing design work for petroleum or chemical companies; consequently, this product has only a vertical demand from manufacturer to end user. Paint products, on the other hand, have a horizontal and vertical demand; all segments of the construction industry use paint products or services during the construction process.

Construction markets can therefore be best targeted from a particular segment or by geographical location. In the sale of products that are used in all segments of the construction industry, the sales representative can start with a knowledge of potential customers within a given location or geographical area. On the other hand, a sales representative for a specialty chemical process equipment manufacturer whose equipment is used only in the industrial segment of the construction industry will be interested exclusively in industrial manufacturing companies. His or her geographical area may be very extensive.

Researching the Needs of the Market

A dramatic increase in basic and statistical information has taken place in the computer and information age of the '80s and '90s, and in the variety of sources from which this data can be obtained. The novice sales representative can often be overwhelmed by the sources and level of detail available. The list below can assist the construction industry sales representative in locating and targeting specific sales opportunities in his or her construction segment.

- Obtain and update a list of the company's current and past sales.
- Obtain information from any relevant chambers of commerce of towns, cities, and states. Compile a list of contractors, subcontractors, vendors, architects, engineers, and owner-type companies that have needed or may need materials and equipment.
- Read and copy any construction or relevant industrial directories compiled by various state commerce departments.
- Compile a list of companies and firms indicating the nature of the business, the number of employees, annual sales figures, names of people in charge, name and telephone number of purchasing agents, and any additional information that may assist

when making a sales call. Today's personal computers make it easy to compile, update, and maintain sales data. This information can be one of the sales representative's main tools for future selling opportunities.

- Read and copy any relevant construction industry trade registers and directories; add information to sales database.
- Establish a practice of reading trade and construction industry journals and magazines. Two of the most useful magazines are the weekly *Engineering News Record* (which has a section devoted to new construction contracts, names and contacts of architects and engineers doing design work on the project, and the name of the successful bidding organization) and the *Constructioneer* (which contains a good deal of information for the sales representative, including information similar to the *Engineering News Record*).
- Obtain or review a copy of the latest edition of the Rand McNally commercial atlas and market guide.
- Obtain or review a copy of the *Commerce Business Daily* published by the U.S. Government. This publication gives information on government procurement invitations, contract awards, subcontract leads, and foreign business and sales opportunities.
- Obtain or review a copy of the *Blue Book* building and construction book. This data source contains a classified and alphabetical listing of general contractors, subcontractors, architects, engineers, material and equipment dealers, and manufacturers and providers of construction services. This annual book has ten regional editions across the U.S.
- Obtain and review names taken from professional and trade associations related to construction.
- Obtain a copy of local business to business yellow pages.
- Subscribe to "lead source" newspapers and newsletters, such as *Construction Data News* and *Construction Market Data*. These publications provide hundreds of new leads each week and provide a good overview of the residential, institutional, commercial, and transportation construction markets. There are many types of lead source publications available, and many are very useful.
- For sales representatives focusing on the manufacturing and industrial construction marketplace, trade magazines such as *Chemical Business*, *Chemical Engineering*, *Electrical World*, *Mechanical Engineering*, and *Pharmaceutical Engineering* discuss future construction projects and can be a valuable resource.
- Once a good customer is located, obtain from him or her leads for other potential buyers. By using this networking method many contacts can be made.

This general information can be of great value to the sales representative as a starting point in making an analysis of a certain segment or geographical area, and for future prospects and sales. Of course, further study, as well as contact and meetings with potential purchasers, is essential.

The Nature of the Construction Market: The nature of the construction marketplace is always changing. To be successful, the sales representative must continually study his or her assigned marketplace, no matter how many or how few customers he or she has. The sales representative goes about such a study systematically, but only to the extent that it is necessary and beneficial.

Sales representatives should stay abreast of latest technology. They should attend trade shows, training seminars, and technical meetings of engineers, architects, and designers. Only in this way can salespeople judge the many positive and negative forces that may be influencing their segment of the construction industry. The sales representative who knows when customers' needs exist or become acute can time sales calls with maximum effect.

The construction market, like all other markets, has many peculiar characteristics. Project opportunities move from place to place and to the inattentive, they appear to spring up unexpectedly. In the '70s and '80s industry had been moving steadily south away from the large northern industrial states; in the late '80s and early '90s this trend appears to have abated. Sales representatives must watch and monitor such changes and trends closely; they must be able to take advantage of them.

Construction markets are continually being created and satisfied. The sales representative cannot rest on past accomplishments; he or she must constantly strive to collect more data on sales prospects and optimize their sales methods and efforts.

The Relationship Between Distribution and Sales

The distribution of products, goods, and services is one of the main activities of every industrial organization. The term *distribution* in this context means the marketing and selling of the industrial organization's products, goods, and services to its targeted market. Efficent distribution is constructive because it adds value to the product or service. Typically, distribution is achieved through the efforts of salesmanship and sales promotion activities, including advertising efforts.

Many manufacturers would agree that in many ways it is less complicated to manufacture a finished product than it is to sell the product. For example, the selling of a new line of doors or windows throughout the construction marketplace is usually a more difficult undertaking than the design and manufacturing process. Distribution concerns the activities and logistics of getting the product or service from the point of manufacture into the hands of the consumer or eventual end user. The success and future growth potential of any business organization can hinge on the effectiveness of its distribution practices.

There is a familiar saying that can be applied to the process of distribution: "If a man invents a better widget, the world will beat a path to his or her door." The originator of this saying neglected to add,"...if the world is notified, aware, and currently in need of (and can afford) this better widget." For a product to sell successfully in today's marketplace, it must satisfy the quality, delivery, and cost requirements of the buyer. The product must have current and future benefit to the purchaser. There are many manufacturers who can produce high quality products and services but have failed to grow and prosper because of shortcomings in their approach to the distribution function.

One of the greatest challenges faced by management in the industrial sector of the marketplace is not concerned with the specifics of engineering, design, manufacture, fabrication or construction – it is controlling and optimizing distribution costs. Larger profit margins await those organizations that can plan and successfully minimize distribution expenditures. As markets develop and expand, the distribution function becomes more and more important.

The ultimate goal of distribution is to put the purchased product or service in the hands of the buyer or end user within a stipulated time frame. As an article leaves its place of manufacture, its usefulness to the purchaser has not yet been realized. Only when the article is delivered and has been put into service is value realized to the buyer and profit made by the producer.

Distribution brings together the manufacturing facility and the organization that uses the finished article. The more efficient the distribution process, the more efficient will be the flow of products, goods, and services to their eventual end users. The force that drives the distribution process is *selling*. Selling sets the distribution process in motion and maintains it so benefits accrue to both seller and buyer. The sales representative cannot successfully perform his or her job without a clear understanding of the distribution process. He or she must know the channels of movement of the manufactured product to the end user.

There are four basic methods by which products and services are distributed to their ultimate buyers.

- **Direct distribution:** The manufacturer sells his or her completed product or service directly to the buyer.
- **Distribution through contractors or subcontractors:** The manufacturer sells his product or service to the contractor or subcontractor, who installs it in the completed construction project. For example, for the installation of air conditioning units, HVAC contractors purchase a specific manufacturer's product and install and maintain the unit.
- **Distribution through third party organizations:** Such organizations focus their attention exclusively on purchasing from the product manufacturer and selling to the ultimate end user, or acting as an exclusive selling agent for the product manufacturer in accordance with a pre-established sales/distributor business relationship.
- **Distribution through other product suppliers and manufacturers:** These suppliers require products and sub-assemblies that they do not manufacture themselves in order to complete what they fabricate for sale to eventual end users. For example, curtainwall fabricator purchases aluminum in standard profiles from one manufacturer, and glass (fabricated per job specs) from another manufacturer, then fabricates them into a curtainwall.

It is important to remember the functions that architects and consulting engineers have in the distribution process. In many instances, architects and consulting engineers do not actually purchase products or services for their clients. Contractors typically purchase the products and services that are described by the architect or consulting engineer in the completed drawings, plans, and specifications.

Figure 7.1 shows several pathways of distribution from a typical equipment manufacturer to the ultimate end user.

The Importance of an Effective Sales Procedure

From a business or legal viewpoint, the sale of an item may be considered as a meeting of the minds of a seller and a buyer through mutual agreement, by which the ownership of an article is transferred. The result of a sale is an exchange of products, goods, or services for money from the seller to the buyer.

A sale is not a single act but a series of related interactions. Although for large and sophisticated items the sales process can be somewhat difficult and time-consuming, many times the total sales process can be relatively simple.

The importance of a workable sales procedure is illustrated by the fact that the average sales representative spends a very short time in actual face-to-face meetings with prospects or customers. Typically, the average sales representative spends over 80% of his or her work week traveling, waiting for meetings, doing paperwork or making sales calls in the office, or checking on the status of fabrication related to the distribution and delivery of current sales. Because the amount of time spent with each prospective customer is so limited, it is vital that the sales representative be prepared to meet each and every question that is asked and to eliminate doubts or uncertainty that may exist in the prospect's understanding of the sale conditions.

A sales procedure is a system that sales representatives follow in their function of selling products and services. The methods employed will never be fully standardized in a formal hard and fast procedure, because there are so many possible variations in the interests, desires, and needs of each prospect or customer. The day-to-day technological

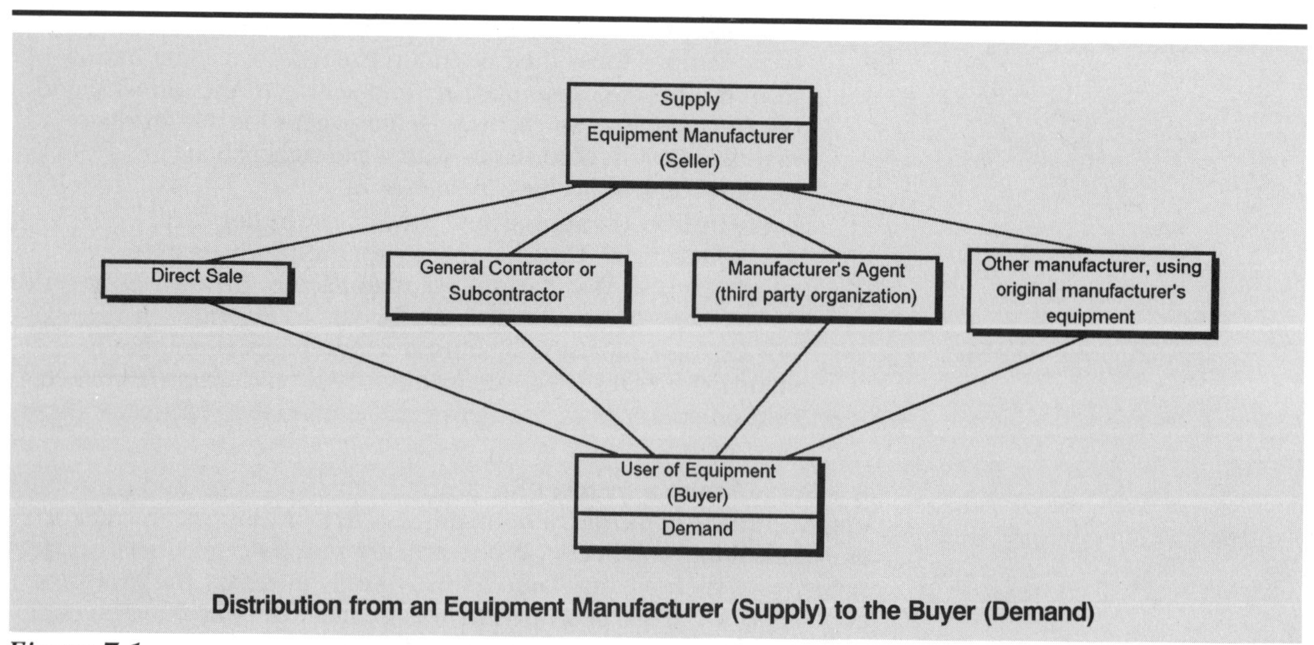

Distribution from an Equipment Manufacturer (Supply) to the Buyer (Demand)

Figure 7.1

advances that are made in the processes of product development and manufacture also make it impossible to standardize the sales procedure. Additionally, the products and services that representatives sell differ dramatically in scope and magnitude, especially in the construction marketplace.

For reasonably standard articles such as doors and windows, the sales procedure may consist only of a single interview or even a telephone call. However, when selling sophisticated equipment, the sales procedure may require many meetings to discuss the many technical and commercial aspects of the future sale. The sales procedure includes a series of interviews, meetings, and presentations along with the necessary offers, quotations, inclusions, exclusions, and any additional paperwork for the buyer in the purchase process.

The Pattern of the Successful Sales Procedure

The process of successfully turning a potential client into a satisfied customer follows a definite pattern. This pattern can be broken down into four general steps:

- **Locate prospective buyers.** The first step in the sales process is to locate the people and organizations that may represent opportunities for future business.
- **Obtain the prospective buyer's interest.** Interest is created in the prospective buyer's mind by the quality, unique features, and capabilities of the article. The quality and effectiveness of the sales presentation also influence the prospective buyer's interest.
- **Encourage the prospective buyer's willingness and desire to do business with the sales representative.** The sales representative must create in the prospective buyer's mind a willingness and desire to do business. This is accomplished by demonstrating that the article being offered is the best of its kind and that it will provide the solution to the prospective buyer's needs.
- **Establish the prospective buyer's respect and confidence.** Respect and confidence can be obtained and fostered only through satisfactory performance – after the article is received and placed into satisfactory operation in the buyer's organization.

Good planning and preparation is necessary; uncertainty, hesitation, and poor planning hamper the effectiveness of the sales procedure. When the sales representative gains the full confidence and respect of the customer and maintains it over the course of time, the difficulty of future selling to that particular customer is reduced substantially.

The Weekly Sales Routine

The following outline is an overall weekly sales procedure or routine that specifies certain goals and targets for the week. This and similar programs have been used very successfully by sales representatives working in the construction marketplace. It may be used as is or modified to meet individual sales situations and requirements.

Weekly Sales Routine

At the beginning of each week, list the names of a set number of people that should be contacted during the week. The list should reflect a combination of prospects and existing customers.

Sales interviews:

Sales presentations with brochures, charts, and diagrams (15-30 minutes per meeting).	20 per week
New contacts: Face-to-face meetings with potential new buyers.	25 per week
Telephone calls to existing customers for follow-up work and maintaining goodwill.	50 per week
Telephone calls to potential new buyers.	250 per week
Lunch and evening sales interviews and meetings.	5 per week
New names and contacts: Names and addresses of potential new customers to add to existing sales database.	250 per week

Research and information gathering:

Read business newspapers, local newspapers, newsletters, trade magazines and related technical journals.	60 minutes each day
Prepare a sales information file on all major or potential customers. This file could be considered a "scrapbook." Copy information from newpapers, newsletters, and magazines. Keep this information current.	15-30 minutes each day

The Tools of Construction Sales

Though a pleasing personality is and always will be one of the most important qualities of a successful sales representative, the representative who relies on personality alone will not be successful. The salesperson must have a full understanding and knowledge of the product or service he or she is representing. In addition to these elements, the sales representative needs to have a sales "tool kit" of items that will compliment his or her knowledge and personality.

Some of these tools are physical aids and some are general selling ideas or themes. Listed below are some of the main tools and promotional materials that a construction-related salesperson might consider.

- Product brochures/descriptive literature
- Technical information on the product or article
- Photographs illustrating the product and showing operational usage
- Videos showing the product being installed and in operation
- Reprints of related technical articles
- Sales flyers describing merits of product

- Calendars, diaries, rulers, key chains, pens, and so on, listing name and address of product manufacturers (and, in some cases, the name and address of the salesperson). These promotional articles should be distributed to current and potential customers.
- Business cards. Professional-looking business cards can produce good results, setting the stage for future sales opportunities.

The most common mistake a sales representative can make is to rely too much on promotional materials. Most sales happen after face-to-face meetings. Promotional materials are tools that the sales representative can use during these meetings, but without the face-to-face contact, they usually will not be successful in producing a sale.

Demonstration

If the product or service can be either easily carried or scaled down to a version that is easily carried, the salesperson should use it to illustrate the product's performance to the potential customer. A demonstration is one of the most powerful selling tools available. When a sales representative successfully demonstrates what he or she is selling, the sales procedure is usually enhanced many times over.

With large specialty equipment, it is usually impossible to perform a demonstration. In this situation, a video or photographs might be used, or a small sub-assembly of the equipment could be used in the sales presentation. When the demonstration has been concluded, the sample should be properly repackaged and set aside. In this way it will not distract the potential customer from continuing the sales interview.

Advertising

One of the most important selling tools is advertising. Advertising assists in establishing and presenting the attributes and characteristics of the product or service being sold by showing what the product can offer the buyer. Advertising helps to mold the opinions of buyers. It also focuses the buyer's attention on the choices available. Listed below are some of the most common forms of advertising used by sales representatives in the construction industry.

- Direct mail advertising: pamphlets, brochures, and flyers
- Advertisements in professional or construction industry magazines
- Advertisements in various trade directories and listings
- Advertising in yellow page telephone directories
- Advertisements in national and local newspapers
- Advertisements on national and local TV/radio stations
- Sponsorship of local sports teams
- Sponsorship of local charities

Building or Facility Openings

Salespeople should never overlook the valuable opportunity of the opening of a building or facility in which his or her products have been used. The salesperson should ensure that his or her organization's name and other relevant details are mentioned in the list of organizations involved in the successful construction of the building or facility.

Testimonials and References

Testimonials and references from satisfied customers are excellent tools for future sales work. Prospective clients are more likely to believe the statements of satisfied customers than the claims of the sales representative. Proof that someone else has used the product or service and is satisfied allays many of the fears that the potential purchaser may have.

Official letterheads and signatures of individuals who were responsible for the purchase give credibility and authenticity to the testimonial. The sales representative should copy four or five of his or her best testimonials and distribute them to potential customers at the appropriate time during sales interviews.

Trade Shows and Professional Society Meetings

Establishing a sales booth at a trade show or a professional society meeting can be an effective sales tool. The booth should be stocked with applicable promotional materials and, possibly, product samples. This approach can gain the undivided attention of potential customers. Customers can ask questions and any potential problems can be discussed on the spot.

To maintain a successful booth, the sales representative must be well prepared. It is usually beneficial to have a technical professional at the booth as well to address specific technical questions from potential customers.

Visits to Engineers and Architects

The sales representative can find it very valuable to visit consulting engineering and architectural firms. These individuals know what is happening in the marketplace. They are aware of any building or plant expansion design and construction work in their market area. Architects and engineers can supply the sales representatives with much useful information.

Written Communication and Quotations

Each business letter, quotation, proposal, and contract that the sales professional sends to future prospects and existing customers is a potential and important sales tool. Because they can either hinder or advance the sales process, all written communications should be used to build confidence in the buyer's mind.

The written communication should have a professional quality that compliments the selling effort. Typically these forms of communication circulate around the buyer's purchasing and business organization, and in many situations they are the only contact between the sales representative and the buyer. These communications should impart to the buyer a professional image of the sales representative and of the product being offered.

The desirable features of a sales representative's business letters, quotations, proposals, and contracts are listed below.

- The written communication should be complete and concise; unnecessary words and thoughts should not be used. Make every word count.
- The paperwork should be neat and attractive. The spelling and grammar must be correct.
- The format and subject matter should be logical and in correct sequence.

- If price is mentioned in the written communication, point of sale, discounts, transportation costs, and any applicable taxes should be described along with any exclusions associated with the price.
- Terms of payment should be described together with a statement of how long the offer is valid.
- Delivery time and transportation methods to be used should be described.

The Sales Interview

When the sales representative has gained as much knowledge as possible about the needs of a prospective buyer, a sales interview should be arranged. Typically the sales interview is carried out at the customer's office; in some cases, it can take place at the sales representative's business office.

As much as the salesperson might want to make the sale during the initial interview, in the vast majority of situations this is not possible. The products or services he or she is representing are usually quite costly. In addition, the buyer usually intends to (or is required by his or her manager to) review other products. Decisions are usually made carefully, often after other products have been considered. The buyer has to consider the cost and operating advantages and disadvantages of each product. The sales representative may have to meet with several individuals associated with the buyer's organization, who will also evaluate the product. The whole process takes time and effort by the salesperson and can take anywhere from one to four weeks.

Planning for the Sales Interview: The prospective buyer will usually select the time for the sales interview. When the salesperson is able to set the interview time, he or she should remember that buyers, just like most business people, like to clear their desks and take care of the previous day's unfinished business in the first hour or two of each business day. The best time to conduct a sales interview is usually between 10:00 a.m. and 4:00 p.m. Remember, lunch time meetings are usually productive and a good setting for discussing and closing sales.

Every sales interview should be well planned. It is a good practice to make a list of points that should be made to the prospective buyer. When traveling to the prospective buyer's office, there is usually time to go over this list mentally and to form a clear image of the sales approach that will be used. Remember that face-to-face contact with potential customers is an important but small part of the sales process. All sales interviews, as well as any other contacts, present an opportunity to strengthen current relationships, create goodwill, and to develop new relationships.

The Content of the Sales Interview: The first step in the initial sales interview is usually the most difficult; first impressions do count. The inexperienced or new sales representative usually thinks long and hard about how to break the ice. After an exchange of introductions, business cards, and pleasantries, the best approach is to get down to business. The sales representative's opening comments should concern the product or service he or she is promoting or selling.

The sales representative should describe the merits of the product or service he or she is selling. Product brochures, sales literature, and any additional information that might be useful to the prospective buyer should be discussed and handed over. The progress of the sales interview will depend on the genuine interest and real needs of the prospective buyer.

If the interview is conducted at the construction site, the sales representative's approach should be as practical as possible. He or she should be prepared to walk around the construction site and get mud on his or her shoes. Salespeople who regularly visit construction sites should invest in their own hard hats, safety shoes, and safety glasses to carry to sales interviews.

Construction superintendents and foremen are sometimes critical of office-based personnel. To overcome this criticism, sales representatives should try to impart that they are prepared to get their hands dirty. The duration of the sales interview should be based on the interest and attention span of the prospective buyer. If the prospective buyer is busy and interrupted often by other individuals or business calls, it may be prudent to arrange for an interview for another time.

When talking in a business environment, many individuals fail to speak clearly; much of the conversation may be poorly punctuated and delivered. This may be the case especially when the sales interview takes place in a setting as unstructured as a construction site. In difficult or stressful meetings, voices often become strained or hard to understand. The speaker or listener may become distracted during the sales interview. Listed below are some basic sales interview do's and don'ts associated with construction selling.

Do:

- Be punctual for any meeting; try to be five minutes early for any appointment.
- Maintain a pleasant and businesslike manner. Exchange business cards.
- Make eye contact with the individual(s) you are talking to.
- Be a good listener. Take notes of the points being discussed.
- Display genuine interest and enthusiasm for the product being promoted.
- Follow each sales interview with a polite letter confirming both the main points of the meeting and the next step in the sales process.

Don't:

- Miss or be late for any appointment.
- Distract or interrupt the prospective buyer while he or she is talking.
- Nervously drum fingers or fiddle with pens or pencils, paper clips, and so on.
- Sit or stand in a slumped or uncomfortable position.

The sales representative should read books or articles or attend seminars about public speaking or sales interview techniques. These publications and presentations can assist in improving future sales interviews.

Discussing Technical Details: When a sales interview focuses on technical matters and functions, one of the most common errors the sales representative can make is to try to discuss detailed technical requirements and terms that are unfamiliar to the potential buyer. No one likes to have his or her lack of experience or knowledge brought out in a business setting.

When speaking with a potential buyer who is unfamiliar with complicated and technical matters, the experienced sales representative will endeavor to use simple illustrations and examples to translate complicated engineering principles to a more practical level of understanding.

The Importance of Profit

There is an old saying that is just as relevant today as it was years ago: "Profits attract investment and growth, losses repel investment and growth." In today's business environment, the importance of profit cannot be overemphasized. Every transaction performed by the salesperson must be profitable and beneficial to the selling organization. The term *profit* has a few basic definitions:

- Profit is the actual monetary gain from any transaction.
- Profit is the gain from business activity; the funds left over when the cost of carrying out the business is subtracted from the money taken in.
- Profit is an advantage or benefit that is realized after concluding an exchange or sale of a particular article.

Business leaders have several goals that must be achieved if the business is to flourish and grow. One of these goals is to generate profits that can be used by the business or its owners. These profits are often reinvested into the business to finance its growth. Other business goals include the following:

- To serve and profit those individuals who invest in the business.
- To serve those individuals that buy products, goods, and services from the business.
- To serve and benefit the community and general public.
- To serve those individuals who are employed by the business.

To fulfill the above goals and objectives and to prosper and grow in the future, the business enterprise must consistently generate profits from its operating revenues. This concept is the foundation that supports the growth and progress of the free market enterprise system. It is the keystone that has contributed to America's amazing growth during the 18th, 19th, and 20th centuries. It is this concept that is being embraced by emerging free market nations around the world.

There are typically four objectives that should be realized and satisfied when disposing of or investing after-tax profits:

- Distribute a realistic percentage of the profit to the owners and shareholders of the business enterprise.
- Distribute a realistic percentage to future research and development of products, goods, and services. (This expense could be considered an on-going business expense.)
- Distribute a realistic percentage to the purchase of manufacturing equipment and machinery, new buildings and manufacturing facilities, and any required transportation equipment.
- Distribute a realistic percentage to outside investments; create a savings fund that will assist the enterprise in future growth opportunities such as the purchase of rival business organizations or other opportunities. This savings fund can also act as an insurance policy for any unprofitable or lean business years.

Profit for the Seller and the Buyer

The sales representative must recognize the impact he or she makes in generating profits for the business. In fact, he or she should help to ensure profits for both the selling organization and the buyer. One of the most important benefits the sales professional brings to the customer is reduced operating costs and increased profits through the use of the products or services that the sales representative provides.

Proposals and Contracts

When preparing a formal proposal or contract for a customer, the salesperson can use a combination of the information that has been discussed and collected during the sales interviews and meetings. The sales representative's proposal or contract should include the following:

- A clear and comprehensive definition and itemized listing of what products or services are to be provided to the buyer.
- Itemized technical specifications, attachments, and drawings together with a list of any revisions or modifications.
- Correct catalog numbers or other product identification numbers or cost codes.
- A clear description and definition of any guaranteed technical or operating performance tests of the product; conditions, times, and terms under which these tests are to be performed.
- Clear statement regarding the responsibility for installation and erection of the equipment or product.
- A credit history review on the potential purchaser.
- Stipulation of the correct purchase price as well as any discounts.
- The delivery date along with a statement specifying whether partial or complete shipments will be made.
- Detailed delivery and shipping instructions, including addresses and contact names.
- Methods or terms of payment together with a statement of how long the offer is valid (typically 30, 60, or 90 days).
- Any applicable taxes, duties, or tariffs; mention of which organization is responsible for payment of such taxes.
- Any retention, cancellation, or late payment charges.
- Correct names, titles, addresses, reference numbers, tax identification numbers, and so on.
- Proper execution and signatures by the appropriate individual(s).

Standard Forms and Contracts

A legal contract under common law is the end product of an offer and subsequent acceptance of that offer. In today's business world, contracts for the selling and purchasing of products and services are often executed by the use of preprinted contracts and purchase orders that itemize the terms and conditions of the sale. Almost every business organization, big or small, has established its own standard contract and purchase order terms and conditions. These forms are usually prepared by lawyers, and their purpose is to provide the maximum legal protection to each business organization.

The number of forms, just like the number of business organizations, can vary considerably. Some business organizations have very complicated and comprehensive forms; others have much less sophisticated forms. When formalizing a sale, the salesperson can usually choose one of the following options:

- Send a preprinted contract or purchase order to the buyer for his or her signature.
- Sign and return the buyer's preprinted contract or purchase order, if the buyer's terms and conditions are acceptable to the sales representative.
- Modify by adding or subtracting from the buyer's preprinted contract or purchase order and send it back to the buyer for signature.
- Develop a compromise contract or purchase order using conditions acceptable to both the seller and the buyer.

The sales representative should always endeavor to protect his or her organization from any legal disputes. When a problem arises concerning which form to use, or when modifications to standard terms and conditions are necessary, the sales representative should seek legal counsel. He or she should always try to negotiate the best terms and conditions available. If the buyer gets hung up with a particular clause or item in the terms and conditions, the sales representative, with assistance from legal counsel, should endeavor to negotiate a compromise that will protect both parties and allow the sale to proceed.

Training and Developing Selling Skills

Every year thousands of young men and women start sales work after completing their formal education. A few of these new sales representatives are successful from the start; however, many fall by the wayside after a short time due to a lack of knowledge and understanding of their industry and market segment. To overcome this problem, a careful plan for training and developing new sales representatives must be established.

Listed below are some of the main approaches that are used by construction organizations to develop the skills of their new or inexperienced sales representatives.

- Send new sales representatives to appropriate training and educational seminars, or purchase sales or marketing videotapes for personnel to view.
- Bring all of the sales representatives together every four to eight weeks for coordination and information sharing. This will help the new sales representatives to become familiar with current industry trends and market conditions.
- Assist the sales representative with any information related to new products the company may be considering.
- Provide monthly or bimonthly training classes that teach salesmanship basics to new sales personnel.
- Allow the new salespeople to accompany more experienced salespeople on sales calls to new or existing clients. This will help the new sales representative to learn by observation and experience.
- Produce and provide to all the sales representatives a sales manual. The sales manual should cover sales procedures and company policy; industries served; geographical territories; standard proposals, contracts, and forms; sales brochures; and other appropriate items. The manual should furnish information that meets the needs of the sales representatives in their daily work effort.

- Inform the sales representative of any special customer or market information, including special situations related to industry expansion or consolidation.

An important factor in any training and development program is the trainer, who must possess practical real life experience, sales knowledge, and teaching ability. He or she must be dynamic and resolute enough to teach the basic points of the training program. Without creating friction, the instructor must be tactful in guiding his or her students.

Overcoming Objections and Obstacles

Motivation, anticipation, and the ability to solve problems are essential skills for the sales representative. The sales representative must strive to understand the customer's thought process. Unanticipated objections and obstacles are barriers in the sales representative's path. If a salesperson does not fully respond to a customer's questions or objections, his or her selling opportunity can be severely compromised.

Answering questions and objections promptly, convincingly, and correctly enhances the sales representative's position and standing. Allowed to stand, problems tend to become magnified and can have a serious impact on the selling opportunity. The sales representative should endeavor to develop a plan for overcoming or solving potential objections and obstacles. Research, information, and specific facts are important tools for countering such objections and obstacles.

The following are some of the most common objections and obstacles that sales representatives encounter during their day-to-day selling efforts.

- The customer's expression of confidence and satisfaction with another supply source.
- Potential problems related to sales conditions: price, commercial terms, technical objections, and delivery time or the country of manufacture may be obstacles to selling a particular product.
- The customer's unawareness of the need for the product. In many cases the customer may need the product, but he or she has not been made aware of the need. The sales representative should describe the merits of purchasing the product.
- The issue of personal preference: "We prefer to buy this particular product from ABC, Inc., because we have always done it this way." The potential customer may have objections about doing business with a new organization.
- Reciprocal purchasing practices: "You scratch my back, I'll scratch yours." In this situation the purchaser and existing seller have established a close business connection that may be difficult for a new organization to break into.

The sales representative should be prepared to sincerely answer any objections brought up by potential customers. The research that the sales representative carries out in preparation for objections and obstacles will educate and strengthen him or her for future sales opportunities.

In considering how best to overcome these objections and obstacles, the sales representative must bear in mind that quality and value are images created in each buyer's mind. The sales representative should base most arguments on the fact that no buyer can be 100% satisfied with all the products, goods, and services that he or she has purchased in the past. There will always be room for improvement.

Overcoming objections and obstacles requires good analytical skills to determine the cause, whether the objection is real or false. Overcoming objections can often be accomplished by pointing out that all organizations are interested in reducing losses, maximizing profits, and increasing productivity. No intelligent buyer can argue that he or she is not interested in saving time and money. When discussing standardization of equipment (i.e., single source supplies) the argument should be, "What if the current supplier goes out of business or stops producing a certain item?" All businesses should have at least two or three sources of supply.

When the sales representative is confronted by competitors' lower prices, the best response is to discuss the issues of quality, reliability, and operation life. Some of these points can deflect the price objection. When a potential customer states that he or she cannot afford a particular product, the salesperson should cite examples of other organizations that have made the investment and list the advantages and benefits that these organizations have realized by using the product.

Sales representatives cannot be successful unless they have an insight into what the buyer is really thinking. By skill, research, and intuition, the sales representative must overcome objections and obstacles. Information should be used in a diplomatic manner that will ultimately break down the objections and obstacles. The end result will be a profitable sale to a new or existing client.

Lost Sales

In analyzing a lost sales order, the sales representative can usually pinpoint some error or mistake made during the sales presentation or in the sales proposal. Before closing the file on a lost sale, the sales representative should list the mistakes made. This list will assist the sales representative when dealing with this particular customer in the future. The most common causes for lost sales are:

- Inadequate research during the early phase of the negotiation process.
- A delayed start by the sales representative.
- A badly planned sales strategy, one that lacks correct details and facts.
- Failure to contact all of the key individuals who influence the purchasing process.
- Failure to understand customer's needs and requirements.
- A lack of knowledge of the product or service being offered.
- A lack of information about competitors' products and services.
- Difficulties with products and services previously sold to the customer.
- Failure to use all necessary sales tools available.
- Lack of initiative and business-related skills in outlining sales details and information to the customer.
- Failure to contribute ideas and solutions to the customer's needs.
- Inadequate understanding, patience, and diplomacy during sales meetings.
- An unprofessional and unconvincing sales presentation that does not meet customer's requirements.
- Badly written and formatted proposals that do not meet customer's needs.

It is always prudent to follow up on a competitor's sales order to ascertain exactly how the competitor fared on the manufacture, installation, and performance of the product or service. If the buyer expresses any complaints, the sales representative can use this information as a tool for future sales opportunities.

Complaints and Dissatisfaction

The sales representative will always encounter dissatisfaction no matter how well the product or services are engineered, designed, manufactured, installed, and performed. Problems may be real or the result of the buyer's unfamiliarity with the new product or service. In selling to customers who are unfamiliar with the product or service, the sales representative should carefully explain its functions, characteristics, and operating methods. Operating manuals should be described and explained. This procedure will usually minimize any start-up or operating problems.

In answering complaints and correcting any problems promptly, the sales representative can demonstrate his or her professionalism and genuine interest in the customer's needs. Ignoring complaints can quickly dispel any goodwill that has been established between the sales representative and the purchaser. Acting quickly and effectively will often strengthen the relationship.

When the problem or complaint has been resolved, the sales representative should check on a weekly or biweekly basis to see that the corrective action has satisfactorily resolved the problem. The sales representative can use this action as a sales tool in obtaining and collecting information and intelligence on new business opportunities.

Checklist of Ideas for Successfully Closing Sales

If ten architects were to design a building, we could almost guarantee that the end product would be ten different designs. Some of the architects would use mainly wood, some stone, and others brickwork. All of the designs would probably serve and satisfy the needs of the potential building owner. Similarly, every sales representative has his or her own unique way to persuade and sell particular products and services to their customers.

Some successful sales representatives are polite, courteous, and low key. Some are persistent and overbearing. How can both types be successful? The answer is that although they demonstrate different personalities and selling approaches, they both share certain key characteristics. They are usually very knowledgeable about their particular product or service. They both want to make sales. They are both energetic. They are both optimistic and enthusiastic. They are both persistent, and finally, they are not easily discouraged.

Listed below are some practical suggestions that may assist sales representatives in their future selling endeavors.

- Successful sales representatives look at their job as a challenge. They are not "turned off" by setbacks; they understand that good days will return if they keep pushing ahead with a positive attitude.
- Courage is one of the best attributes a sales representative can have. With it the sales representative can move ahead to victory; without it the battle is lost before it even begins.

- Fear is a natural feeling. Without it the sales representative might overlook some vital detail, but he or she cannot allow fear to overcome victory.
- A sales representative must be sincere and believe 100% in what he or she is promoting and selling.
- Enthusiasm is a trait common to all of the most successful sales representatives. It is the force that makes sales happen. Enthusiasm grows out of knowledge and confidence.
- One of the main attributes of a sales representative is persistence. Nothing in the world can take the place of this quality. Knowledge, ability, and talent cannot overcome dogged and resolute persistence.
- Optimism is a vital trait to all salespeople. The sales representative cannot expect every sales lead to turn into an order; he or she should view failure or rejection as a learning experience. The sales representative can turn present defeat into future victory.
- Salespeople should imitate and adopt attitudes, mannerisms, methods, procedures, and habits of other successful sales representatives.
- The sales professional should continue to pursue a sales opportunity even if the original plan did not work. Think of an alternative plan that will succeed and press on.
- One of the greatest qualities a sales representative can have is courtesy. It does not cost anything to be polite. The gains, on the other hand, can be considerable.
- Pride can be a plus or a minus to a sales representative. In its extreme form pride can be overbearing. When used positively, it can be a great motivating force. Pride in what he or she sells can be a great strength to a salesperson.
- The sales representative should know facts and details about the organization he or she works for. It can be useful to compile a listing of facts and information that makes the organization special.
- Successful sales professionals read books and articles about positive thinking and self improvement. These books and articles can really help in motivating individuals.
- Many people are quick to criticize other individuals. This is a negative attitude. Sales representatives should always look for the good points in all individuals; this approach is positive and will help create goodwill.
- Salespeople should smile and have a pleasant demeanor. People feel more comfortable doing business with friendly and pleasant individuals.
- The sales representative should forget past defeats. Thinking and dwelling upon them dilutes and weakens the effort required for the next selling campaign.
- Participating in local community activities, (e.g., school board, Girl Scouts, Boy Scouts, Girl Guides, church activities, local sports activities, etc.) presents individuals with opportunities for meeting and making many acquaintances and friends. This assists the sales representative with future networking opportunities.

- When opening a conversation with a potential customer, the salesperson should ask questions and look and listen for signs on which matters are of interest to the customer. The easiest way to create a relationship is to have an interest in a common subject.
- All successful sales representatives have something in common; they are hard working. Hard work is one of the main ingredients for success.
- Tact is a quality that sales representatives require. It is the ability to understand others and to be broadminded and generous in dealing with individuals and their shortcomings. Tact and diplomacy generally go a long way in overcoming hostility, opposition, and difficult individuals.
- There is nothing more important to a sales representative than his or her list of possible future customers. A sales representative without future prospects is in a destitute condition.
- Many new sales representatives tend to get overwhelmed by the scope and magnitude of the sales process and procedure. The secret of overcoming big tasks is to break the task down into small and more manageable work units.

A group of experienced sales managers were asked to list the ten most important qualities that they had observed in their most successful sales representatives. The group, after much discussion, produced the following list:

- Knowledge: A detailed understanding of the product being sold and knowledge of the competition
- Courage: the will to succeed; not taking no for an answer
- Ambition: the strong desire to improve one's lot or station in life
- Confidence: Belief in the product one sells and in one's ability to sell
- Energy: the vigor, strength, and concentration to perform one's job at an optimized level
- Courtesy: being considerate and thoughtful of others
- Conversation skills: the ability to converse and talk with others
- Sincerity: the ability to be honest and free of pretense or deceit
- Good appearance: the facility to be well dressed and groomed
- Memory: the ability to retain relevant information that will assist in future selling opportunities

The Ethics of Salesmanship

The ethics and professional guidelines that sales representatives work within are usually established by the sales manager. Although most businesses have similiar ethical guidelines, there are usually some differences in each business's views toward conduct, duty, and judgment. Sales representatives must realize that they cannot act without regard for their fellow salespeople and competitors. Competition is healthy and will always be ardent and fierce. However, the sales representative must always compete in an honorable, clean, and proper manner without harming or wronging any other individual or business. The following is a list of considerations for formulating a code of ethics related to construction selling activities.

- Take no unfair advantage of a person's lack of knowledge or experience. Be truthful and straightforward. Do not make false statements or claims to current or potential customers.

- Conduct yourself in an honest and straightforward manner with both customers and competitors.
- Assist new sales personnel when they are joining the organization. Help them by providing information and assistance in overcoming any difficulties they may encounter during the first months of their new assignment.
- Keep your word. If you have made a promise to a customer regarding price or delivery of an article, go out of your way to ensure that the promise is kept.
- Deal with people as you would like to be dealt with: with honesty, integrity, and respect.
- Maintain a broad and balanced outlook: realize the merit in the ideas and views of other individuals.
- Carefully guard your reputation and display good moral character and judgment.

Appendix, Glossary and Index

Appendix A

This section contains 54 forms, including examples of purchasing administration forms, bid tabulations and analyses, change orders, status reports, and record logs. Many of the forms and reports have been used as examples throughout this book. They can be adapted for specific applications.

Purchase Requisition

Project No. ______
Project Name______
Project Location ______
Date______
Page 1 of ______
Prepared By______

To: ______

From: ______

Subject ______

Purchase Requisition No.______
Dated ______

☐ Purchase ☐ Rental
☐ Lease ☐ Service Contract
☐ Maintenance ☐ Other______

Delivery Required By ______
Delivered To______

Attachments ______

☐ Sales Tax Included
Weight ______
☐ Drawings Required
☐ Freight Included

Inspection Required
☐ No
☐ Yes By______

Reference No.	Quantity	Description of Items to be Supplied	Unit Price	Total
			Total	**$**

Promised Delivery Date ______ Ship Via ______

Remarks ______

Distribution ______ Action By ______

Purchase Order

(This is an offer to purchase on the terms and conditions stated on the reverse side of this document)

Project No. ______
Project Name______
Project Location ______
Date______
Page 1 of ______
Prepared By______

Purchase Order No.

To: ______

Ship To: ______

Item No.	Quantity	Description	Code or Equip. No.	Unit Price	Extension Value

Confirming Order Placed With

Total Value **$**

Cost Code

Payment Terms

Invoice Instructions
Send ☐ (_____) copies of invoice with original bill of lading to:

Shipper

☐ Sales Tax Included

Ship Via

Weight/Lbs.

Cube/CF

Shipping Terms

Promised Shipment

The Attachments Below Are Hereby Made a Part of This Contract

☐ General Terms and Conditions ______

☐ Drawings ______

☐ Specifications ______

☐ Vendor's Proposal______

☐ Other ______

Approval By

Title

Company Date

Seller hereby accepts and agrees to all terms and conditions, and only those terms and conditions contained on the reverse side of this purchase order.

Accepted by

Title

Company Date

Distribution ______

Action ______

Purchase Order Continuation Sheet

Project No. ____________________
Project Name____________________
Project Location ____________________
Date ____________________
Page_______ of ____________________
Prepared By____________________
Revision No. ____________________

Purchase Order No.

Item	Quantity	General Description of Materials/Services	Unit Price	Extension
			Total	$

Purchase Order Log

Project No. ____________________
Project Name ____________________
Report No. ____________________
Date ____________________
Page __________ of __________

P.O. Number	P.O. Date	Revision No.	Vendor/Supplier Name	Description	$ Value	Delivery Date

Comments: ____

Purchase Order Change Order

Project No. ______________________
Project Name______________________
Purchase Order No. ________________
Change Order No. _________________
Page ______ of ___________________
Date _____________________________

To:	Telephone No.
	Fax No.
Attn:	
Confirmed to:	Date

Details of Change Order:

Item No.	Quantity	General Description of Materials/Services	$ Extension
		Total Value of Change Order	**$**

Original purchase order amount $ ____________

This change order value $ ____________

Revised purchase order amount $ ____________

Effects delivery by ________ days (+ or –)
(add or subtract as necessary)

Soder Construction and Development Corp.
Approval Authority

By ______________________

Title ______________ Date ______________

Seller hereby accepts and agrees to all terms and conditions of Purchase Order No. ____________

Dated ________________ and this Change Order.

Accepted by ______________ Title ______________

Company ______________ Date ______________

Bid Opening Sheet

Scope of Work: ______________________

Project No. ______________________
Location ______________________
Compiled By ______________________
Witnessed By ______________________
Page ________ of ________
Date ______________________

Construction Bid Package No.	Bidders Name & Address	Remarks/ Comments	Date Out For Bids	Bid Due Date	Date Bid Was Rec'd	Time Bid Was Rec'd	Bid Opening Date	Total Bid Value

Remarks ______________________

Equipment Bid Summary

Project No. ______________________
Project Location ______________________
Date ______________________
Compiled by ______________________
Date Required at Site ______________________
Cost Code ______________________
Page ____________ of ____________

Vendor/Supplier			1	2	3	4	5
Country of Origin							
Bid Reference							
Bid Date							
Period of Validity							
Bid Currency							
Project Exchange Rate							
Item	**Qty**	**Description**	**Bid**	**Bid**	**Bid**	**Bid**	**Bid**
Quoted price ex-works							
Discounts							
Inspection-testing							
Packing-export prep.							
FOB charges							
Insurance costs							
Est. transport cost to site							
Customs, duty tax, tariffs, etc.							
Other costs							
Est. total cost on site							
Quoted delivery ex-works							
Est. transport time							
Est. total delivery to site (date)							
Payment terms							
Special terms							
Total Evaluated Cost							

Recommended by Purchasing Agent

Reasons: Lowest Price ☐ Acceptable Delivery ☐ (tick for yes)

Remarks:

Purchasing agent Date

Recommended by Engineer

Reason: meets technical requirements ☐ (tick for yes)

Remarks:

Engineer Date

Recommended by Project Manager

Reason: meets price/delivery/project objectives ☐ (tick for yes)

Remarks:

Project manager Date

Commercial Bid Analysis

Contract Work Scope ______________________

Contract Cost Code ______________________
(Unit Price, Lump Sum, Cost Plus, etc.) ______________________

Project No. ____________
Project Name____________
Location ____________
Client ____________
Date____________
Page ______ of ______

Description	CONTRACTOR PRICING DETAILS						
	1 Contractor	2 Contractor	3 Contractor	4 Contractor	5 Contractor	6 Contractor	Budget Estimate
Total all-inclusive price							
Price adjustments made due to the exceptions taken to the specifications and/or conditions of contract by contractor/subcontractor.							
Price adjustments made due to (A) ______							
Price adjustments made due to (B) ______							
Price adjustments made due to (C) ______							
Price adjustments made due to (D) ______							
Price adjustments made resulting from letters of clarification and faxes.							
Price used for comparison/evaluation (lump sum or unit price contracts)							
Selected Bid							

Remarks: ______________________

Signatures of Approval:

______________________ Purchasing Dept. (Name/Title/Date)

______________________ Project Mgr. (Name/Title/Date)

______________________ Client Representative (Name/Title/Date)

Bid Tabulation & Purchase /Authorization

Recommended Vendor ______________________
____
Reason Recommended ______________________
____

Project No. ______________
Project Name ______________
Date ______________
Page ________ of ________
Prepared By ______________

Budget: $ ________		Contract/Scope of Work: ________ ________	Bid No. 1 S/C Name	Bid No. 2 S/C Name	Bid No. 3 S/C Name	Bid No. 4 S/C Name	Bid No. 5 S/C Name
Item No.	Quant.	Description					

Approval	Date		Quoted Cost Totals					
Purch.			Firm Price					
Proj. Engr.			Est. Adjust.					
Proj. Mgr.			**Total Evaluated Cost**					
Bus. Mgr.			Bond Costs					
Vice Pres.								
Client Approval		Delivery At Jobsite	Req'd					
Other			Quoted Date					
Other			Meets Specs (Yes or No)					

Subcontractor or Vendor Verbal Quotation Record

Project No. ____________
Project Name ____________
Project Location ____________
Date ____________
Page ___ of ____________
Prepared By ____________

Subcontract ☐	Material/Equip. ☐	Services ☐	Verbal Quotation ☐

If Material, FOB Point: ____________	Via Ship ☐	Via Air ☐	Via Truck ☐	Via Rail ☐	Type of Work: ____________

Subcontractor or Vendor:	Phone No.	FAX No.	Date
Address ____________	Name of Person Contacted/Title:		

Equipment/Materials or Services Required

Item No.	Description	Quantity	Unit	Unit Price	Amount

Number of Days that Bid is Valid (________Days) **Total Bid Value $** ________

Vendor Assistance to be Furnished by ____________

Delivery Commencement Date ____________	Date Delivery to be Completed ____________	Delivery to be made at ________Items per ____

Items Missing	Yes	No	N/A	Exclusions/Exceptions
Engineering Costs				
Bonds				
Sales Taxes				
Complete Scope of Work				
Freight				
Escalation				
Price is Firm				

Remarks: ____________

Confirmation Being Sent

By FAX ☐ By Letter ☐

On the Following Date ____________

Quotation Received By: ____________

Date: ____________

Telecon Memo Summary

Project No. ______________
Project Name ______________
Project Location ______________
Date ______________
Page 1 of ______________
Prepared By ______________

From: ______________
To: ______________
Subject: ______________

Reference No. ______________
Telecon Date ______________
Time ______________
Representing ______________
Telephone ______________
Representing ______________
Telephone ______________
Fax No. ______________

Summary/Details: ______________

Remarks: ______________

Signed: ______________ Date: ______________

Copies Forwarded To: ______________

Date: ______________

Distribution: ______________

Action: ______________

Agreement Reached Record

Project No. ____________
Project Name____________
Project Location ____________
Date____________
Page ______ of ____________
Prepared By ____________

Bidder Name/Address

____
____
____
____

(A) Quotation Ref. No. ____________ Dated ____________

☐ is in full compliance with Request for Quotation.

☐ is not in full compliance with Request for Quotation.

(A-1) Technical variations to be considered:

____
____

(A-2) Commercial variations to be considered:

____
____

(B) Required changes that differ from Request for Quotation: ____________

____

(C) A possible contract/purchase order will be based on:

☐ minutes of meeting dated ____________

☐ the Request for Quotation in its entirety, except for:

____

☐ the General Terms and Conditions, except for:

☐ ____________

☐ ____________

(D) The period of warranty is ____________

(E) Date of delivery or completion ____________

(F) Price basis/exceptions ____________

(G) Taxes (state and city) ____________

(H) Packing costs ____________

(I) Freight costs ____________

(J) Unloading costs ____________

Agreement Reached Record (cont'd.)

Bidder Name/Address

____
____
____
____

Project No. _______________
Project Name _______________
Project Location _______________
Date _______________
Page _______ of _______________
Prepared By _______________

Quotation Ref. No. _______________

(K) Penalties/Liquidated damages _______________

(L) Bonds _______________

(M) Adjusted value of bid considering all changes and/or modifications laid down in this document:

(M-1) The new Bid Value is: $ _______________ _______________
Bidder's Initials

(N) Bid prices are firm until: _______________ _______________
Bidder's Initials

(O) Terms of Payment: _______________

(O-1) All Progress Payment advances must be accompanied by an appropriate Waiver of Lien Form. First Progress Payment will be submitted on _______________. Subsequent Progress Payments will be issued thirty days from first Progress Payment submission.

(P) Contract/Purchase Order's "Scope of Work" and its Terms and Conditions have been fully discussed and are understood by Bidder.

☐ Yes _______________

☐ No _______________

Comments: _______________

____

(Q) By signing this page, the Bidder confirms that his or her quotation will stay valid for contract award until

____
(Date)

(R) Other comments: _______________

____

Bidder:

_______________ _______________
(Authorized representative) (Title)

_______________ _______________
(Name and title) (Date)

For the Company:

_______________ _______________
(Company name) (Title)

_______________ _______________
(Authorized representative) (Date)

Client Change Order

Issued To: ______________________

____

____

____

Project No. ______________________
Project Name______________________
Prepared By ______________________
Contract Number ______________________
Change Order No. ______________________
Page __________ of ______________
Date of Issue ______________________

The following change is hereby made a part of your contract with ______________________ which shall be performed under the same terms and conditions as required by the original contract. Except as modified and described below, the original contract and all prior amendments and changes shall remain in full force and effect.

Cost Code Number	**Description of Work**	**+/- Add/Deduct**	**$ Value**

Comments ______________________

Original Contract Price: $ ______________

C.O. #1 thru C.O. _____ (increased) (decreased) (no change) $ ______________

This Change Order: (increases) (decreases) (unchanged) $ ______________

New Total Contract Price Is: $ ______________

Contract duration will be (increased) (decreased) (unchanged) ______________ days

(   ) Delete where necessary

Company Name: ______________________
By: ______________________
Signature: ______________________
Title: ______________________
Date: ______________________

Client Name: ______________________
By: ______________________
Signature: ______________________
Title: ______________________
Date: ______________________

Subcontract Change Order

Issued To: ______________________
____
____
____

Project No. ______________________
Project Name______________________
Prepared By ______________________
Contract Number ______________________
Change Order No. ______________________
Page __________ of ______________
Date of Issue ______________________

The following change is hereby made a part of your contract with ______________________ which shall be performed under the same terms and conditions as required by the original contract. Except as modified and described below, the original contract and all prior amendments and changes shall remain in full force and effect.

Cost Code Number	Description of Work	+/- Add/Deduct	$ Value

Comments ______________________
____
____
____
____
____
____
____
____
____
____
____

Original Contract Price: $ ______________________

C.O. #1 thru C.O. _____ (increased) (decreased) (no change) $ ______________________

This Change Order: (increases) (decreases) (unchanged) $ ______________________

New Total Contract Price Is: $ ______________________

Contract duration will be (increased) (decreased) (unchanged) ______________ days

( ) Delete where necessary

Company Name: ______________________
By: ______________________
Signature: ______________________
Title: ______________________
Date: ______________________

Subcontractor Name: ______________________
By: ______________________
Signature: ______________________
Title: ______________________
Date: ______________________

Emergency Change Order (E.C.O.)

Project No. ____________________
Project Location ____________________
E.C.O. No. ____________________
Page ________ of ____________
Date ____________________

To: ______________________________ From: ______________________________

Attention: ______________________________
Reference: ______________________________

Reason for emergency change order: ______________________________

Description of work: (Include detailed report, statement, or photographs to support E.C.O.) ______________________________

______________________________ ____________
Subcontractor's/Vendor's Representative Date

______________________________ ____________
Site Manager Date

NOTE: The increase in contract sum authorized by this document shall not exceed $ ____________.

The Subcontractor's cost evaluation for this work must be submitted within ________ days from when E.C.O. work was commenced.

Originator of E.C.O.:

______________________________ ____________
Name/Title Date

Remarks: ______________________________

Distribution: ______________________________

Action: ______________________________

Addendum

Project No. ______________________
Project Name ______________________
Addendum No. ______________________
Page ________ of ______________
Date ______________________

Reference: ______________________

Bid Package No. ______________________

To: ______________________

____

Project No. ______________________

Project and Location: ______________________

____

____

Engineer/Architect: ______________________

Description of addendum: ______________________

____

____

____

The above referenced bid package is hereby modified. All provisions and requirements contained within said bid package shall remain in effect unless specifically indicated otherwise.

Specifications:

Specification Section	Specification Title	Specification Revision Number

New Drawings:

Drawing Title	Drawing Number	Drawing Date

Revised Drawings:

Drawing Title	Drawing Number	Drawing Date	Revision Number	Revision Date

Bid due date remains the same ______________________

(or)

Bid due date has been revised to ______________________

For: ______________________
(Company Name) (Name) (Signature) (Date)

Distribution: ______________________

Bulletin

Project No. ______________________
Project Name______________________
Contract No. ______________________
Purchase Order No. ______________________
Bulletin No.______________________
Page__________ of ______________
Date ______________________

To: ______________________

Attn: ______________________

Necessary Action:

This Bulletin covers certain specific changes to the work covered by above referenced Subcontract Purchase Order. The provisions, drawings, and specifications of said Subcontract and previous modifications and changes shall apply unless specifically indicated otherwise.

(A) ☐ The Subcontractor/Vendor is hereby authorized to proceed with this bulletin. Work should proceed immediately.

(B) ☐ The Subcontractor/Vendor is not authorized to proceed with this bulletin until Owner provides written authorization to proceed. If waiting for written authorization causes any delay in the schedule, notify Contractor immediately.

(C) ☐ The Subcontractor/Vendor shall submit, within five days after the receipt of this Bulletin, an itemized cost breakdown of any price reduction or addition and any schedule impact caused by the revisions or additions contained in this bulletin.

I. Scope or description of Bulletin: ______________________

II. Drawings changed or modified by Bulletin: ______________________

III. Specifications changed or modified by Bulletin: ______________________

Company Name: ______________	Accepted by: ______________
By: ______________	Company Name: ______________
Signature: ______________	Signature: ______________
Title: ______________	Title: ______________
Date: ______________	Date: ______________
Distribution: ______________	Action: ______________

Change Order Status Log

Project No. ____________
Project Name____________
Date____________
Page ______of ________
Prepared By ____________

Change Order Number	Emergency Change Order Number	Date Initiated	Description of Change Order	Date Change Order Received	Change Order Submitted		Date Negotiated	Adjusted Change Order Value		Remarks/Current Status
					Date	Amount		Amount	Duration Adjustment	

Contract Modification Report

Contract Number ______________________

Name of Contractor/Subcontractor ______________________

Project No. ______________

Project Name ______________

Page ______ of ______

Date of Report ______________

Change Order No./ Emergency Change Order No.	Description of Change Order	Change Order Date Approved	Change Order Amount Approved	Original Contract Amount $ ______	Original Number of Days ______	Original Completion Date ______
				Current Contract Amount	Number of Days Approved	Current Completion Date

Remarks ____

Inspection/Expediting Status Report

Project No. ____________________
Project Location ____________________
Client ____________________
Purchase Order No. ____________________
Cost Code ____________________
Requisition No. ____________________
Specification No. ____________________
Revision ____________________

This Report No.	Page ____ of ____	Date of this visit	Date of last visit	Last Report No.	Inspector/Expediter/Buyer
________		________	________	________	________

Main Supplier Information

Vendor ____________________
Address ____________________
Vendor Reference No. ____________________
Personnel Contacted ____________________
Telephone No. ____________ FAX No. ____________
Equipment/Package ____________________

P.O. Delivery
Week number ____________
Date ____________

Current Delivery
Week number ____________
Date ____________

Next Visit ____________
To expedite and audit ☐
To inspect/test ☐
To inspect packing ☐
For final inspection ☐

Sub-Vendor Information

Sub-Vendor's Order No. ____________ Date ____________
Sub-Vendor ____________________
Address ____________________
Sub-Vendor's Reference No. ____________________
Personnel Contacted ____________________
Telephone No. ____________ FAX No. ____________
Equipment/Package ____________________

Delivery to Vendor
Week number ____________
Date ____________

Latest Delivery
Week number ____________
Date ____________

Next visit ____________
To expedite and audit ☐
To inspect/test ☐
To inspect packing ☐
For final inspection ☐

Purchase Order Status

Progress ahead of schedule		Slippage to current schedule	Complies with specification/date			Tests witnessed as specified/date			Released for packing/date			Released for shipping/date		
Yes	___ Weeks	___ Weeks	Yes		____	Yes		____	Yes		____	Yes		____
No			No		____	No		____	No		____	No		____

Remarks/Comments: ____________________

Compiled by: ____________________
Name Title Date

Distribution: ____________________

Action: ____________________

Equipment Purchasing Status Report

Project No. ____________
Project Name ____________
Location ____________
Date ____________
Page 1 of ____________
Prepared By ____________

Equipment No.	Description	Requisition No.	Spec to Purch. Schedule/Actual	Out For Bid	Bids Due/Rec'd	Bids to Eng.	Tech. Eval. Comp. Schedule/Actual	Award Date	P.O. No.	Remarks
			(S)		(D)		(S)			
			(A)		(R)		(A)			
			(S)		(D)		(S)			
			(A)		(R)		(A)			
			(S)		(D)		(S)			
			(A)		(R)		(A)			
			(S)		(D)		(S)			
			(A)		(R)		(A)			
			(S)		(D)		(S)			
			(A)		(R)		(A)			
			(S)		(D)		(S)			
			(A)		(R)		(A)			
			(S)		(D)		(S)			
			(A)		(R)		(A)			
			(S)		(D)		(S)			
			(A)		(R)		(A)			
			(S)		(D)		(S)			
			(A)		(R)		(A)			
			(S)		(D)		(S)			
			(A)		(R)		(A)			
			(S)		(D)		(S)			
			(A)		(R)		(A)			
			(S)		(D)		(S)			
			(A)		(R)		(A)			
			(S)		(D)		(S)			
			(A)		(R)		(A)			
			(S)		(D)		(S)			
			(A)		(R)		(A)			
			(S)		(D)		(S)			
			(A)		(R)		(A)			
			(S)		(D)		(S)			
			(A)		(R)		(A)			
			(S)		(D)		(S)			
			(A)		(R)		(A)			
			(S)		(D)		(S)			
			(A)		(R)		(A)			

Purchase Order Status Summary

Project No. ____________
Project Name ____________
Project Location ____________
Date ____________
Page 1 of ____________
Prepared By ____________

P.O. No.	Award Date	Description (Equip. No.'s)	Vendor/ Supplier	P. O. Amount	C.O. No./$	Original Budget	Shop Drawings			Approved Shop Drawings			Shipment			Rec'd At Site	Remarks
							Due	Rec'd	To Eng.	Due	Rec'd	To Vendor	Sched.	Quoted	Actual		

Equipment Delivery Status Report

Project No. ____________
Project Name ____________
Project Location ____________
Date ____________
Page 1 of ____________
Prepared By ____________

P.O. No.	Vendor/ Supplier	Equipment/ Item	P.O. Date	Vendor/ Supplier Location	Shipment Dates				Expedited Date	Remarks/Comments
					P.O. Date	Revised	Current	Actual		

Receiving Report

Project No. ______________________
Project Name ______________________
Project Location ______________________
Date ______________________
Page ______ of ______________________
Prepared By ______________________

Equipment No.: ______________ Equipment Tag No.: ______________ Purchase Order No.: ______________ Bill of Lading No.: ______________

Vendor/Supplier: ______________________________ Contact Name: ______________________

Shipped from: ______________________________ Telephone No.: ______________________

Freight Carrier: ______________________________ FAX No.: ______________________

Received at: ______________ Date: ______________ Gross Weight: __________ lbs. (or) __________ kilos

Condition of Equipment: ____

Damages (if any): ____

____

____

Damage Claim Required: No ☐ Yes ☐

Damage Claim Written: Date: ______________ By: ______________ Ref: ______________

Unloaded By: ______________________ Storage Location: ______________________

Packing Materials to be Returned: No ☐ Yes ☐ Details: ______________________

P.O. Line Item	Equip. No.	Qty. Ordered	Qty. Rec'd	Date Rec'd	Shipment: (C) Complete (P) Partial (FP) Final Partial	Remarks

Received By: ______________________________ Date: ______________________

Representing: ____

Checked By: ____

Distribution: ______________________________ Action By: ______________________

Shipping Report

Project No. __________
Project Name __________
Project Location __________
Date __________
Page 1 of __________
Prepared By __________

Item No.	Contract No./ P.O. No.	Equipment Number/ Tag No.	Value	Weight	Transport Company	Contact Name FAX No. Telephone No.	Port of Loading Contact Name FAX No. Telephone No.	Port of Unloading Contact Name FAX No. Telephone No.	Freight Clearance Date

Remarks/Comments __________

Distribution __________

Action __________

Outgoing Correspondence Record Log

Project No. ______________
Project Name______________
Project Location ______________
Date______________
Page 1 of ______________
Prepared By ______________

Outgoing Corresp. Number	Correspondence To	Subject	Date Sent	File Reference Number	Date Answer Answered	Incoming Corresp. Number	Remarks/ Comments

Incoming Correspondence Record Log

Project No. ________________
Project Name________________
Project Location ________________
Date________________
Page 1 of ________________
Prepared By ________________

Incoming Corresp. Number	From	Subject	Date Received	File Reference Number	Date Answered	Outgoing Corresp. Number	Remarks/ Comments

Potential Claim Report

Project No. ______________________
Project Name______________________
Project Location __________________
Date_____________________________
Page _________ of ________________
Prepared By ______________________

To: ______________________________

From: ____________________________

Subject: __________________________

Contract No.: ______________________

P.O. No.: _________________________

Dated: ____________________________

Report No.: _______________________

The following are a listing of items related to your contract/P.O. that have come to our attention which could result in a potential claim to your organization.

Item No.	Description of Potential Claim Elements
1	Description of Potential Claim: ______________________
2	Background of Potential Claim: ______________________
3	Statements or records related to Potential Claim: ______________________
4	Photographs or additional evidence related to Potential Claim: ______________________
5	Suggested remedial actions: ______________________
6	Additional information: ______________________

Prepared By: ______________________
Title: ______________________
Representing: ______________________
Date: ______________________

Received By: ______________________
Title: ______________________
Representing: ______________________
Date: ______________________

Distribution: ______________________

Action: ______________________

Backcharge Notification

Project No. ______________________
Project Name______________________
Project Location ______________________
Date______________________
Page ________ of ______________
Prepared By ______________________

To: ______________________________

From: ______________________________

Subject: ______________________________

Contract No.: ______________________

P.O. No.: ______________________

Dated: ______________________

Report No.: ______________________

The following is a listing of corrective work that your organization is required to perform related to your contract/P.O.

(A) ______________________________

(B) ______________________________

(C) ______________________________

Item No.	Results of Backcharge Notification
1 ☐	Vendor/Subcontractor will perform corrective work at their own cost.
2 ☐	Vendor/Subcontractor is unable to perform corrective work on a timely basis and will accept backcharge for actual costs incurred.
3 ☐	Vendor/Subcontractor declines to perform corrective work and will not accept backcharge. (See notes below)

Notes: If Item 3 is checked, provide a brief explanation as to the reasons Vendor/Subcontractor declines or objects to backcharge.

Prepared By: ______________________
Title: ______________________
Representing: ______________________
Date: ______________________

Received By: ______________________
Title: ______________________
Representing: ______________________
Date: ______________________

Distribution: ______________________

Action: ______________________

Deficiency Report

Project No. ______________________
Project Name______________________
Project Location ______________________
Date______________________
Page ______ of ______________________
Prepared By ______________________

To: ______________________
____

Contract No.: ______________________

From: ______________________
____

P.O. No.: ______________________

Subject: ______________________
____
____

Dated: ______________________

Report No.: ______________________

The following deficiencies in your contract/P.O. have come to our attention. In accordance with your contract/P.O., you have _________ days to correct these deficiencies.

Item No.	Description of Deficiency
1	
2	
3	
4	
5	
6	

Prepared By:______________________
Title: ______________________
Representing: ______________________
Date: ______________________

Received By:______________________
Title: ______________________
Representing: ______________________
Date: ______________________

Distribution: ______________________
____

Action: ______________________
____

Backcharge Register

Project No. ____________

Project Name ____________

Project Location ____________

Date ____________

Page 1 of ____________

Prepared By ____________

Report No. ____________

B/C No.	Date of B/C	P.O./ Subcon. No.	Vendor/ Subcontractor	Description of Backcharge	Work Order Ref.	Est. Total Value	Est. Recover-able Amt.	Expend. To Date	Total Forecast Amount	Amount Recovered To Date	Remarks/Comments
				Page Total							

Distribution ____________

Transmittal Form

Project No. ______________________
Project Name______________________
Project Location ______________________
Date______________________
Page_______ of ______________
Prepared By ______________________

To: ______________________ Attention: ______________________

From: ______________________

Subject: ______________________

The following is transmitted:

- ☐ Herewith ______________
- ☐ Under separate cover ______________
- ☐ For your use or distribution ______________
- ☐ For review and comments ______________
- ☐ For correction and resubmittal ______________
- ☐ And is approved ______________
- ☐ For your files ______________
- ☐ For your approval ______________
- ☐ For your information ______________
- ☐ Other ______________

Number of Items	Description	Dated	Reference

Remarks ______________________

Distribution: ______________________

Compiled By: ______________________ Date: ______________________

Payment Application/ Requisition

Project No. ____________
Project Name ____________
Project Location ____________
Date ____________
Page ______ of ______
Compiled By ____________

Contractor/Subcontractor ____________

Address/Contact/Telephone ____________

Scope of Work ____________

Application No. ____________

Contract/Subcontract No. ______ Dated ______

For work performed in following period ____________

CSI Division No.	Division Description	Value of Each Division	% of Work Completed		Total Completed To Date
			Previous Application	This Period	
1	General Requirements				
2	Site Work				
3	Concrete				
4	Masonry				
5	Metals				
6	Wood & Plastics				
7	Thermal & Moisture Protection				
8	Doors & Windows				
9	Finishes				
10	Specialties				
11	Equipment				
12	Furnishings				
13	Special Construction				
14	Conveying Systems				
15	Mechanical				
16	Electrical				
Base Contract Total Value					
C.O. No.	Description of C.O.	Value of C.O.			
Approved Change Order Total					

Total Amount of Work Completed to Date (Date ________):	
Less Retainage as per Contract:	
Subtotal Value:	
Less Amount Previously Received by Contractor/Subcontractor:	
Total Amount Due This Application:	

Distribution: ____________

Action: ____________

Insurance Status Report

Project No. ____________________
Project Name____________________
Project Location ____________________
Date____________________
Page ______ of ____________
Compiled By ____________________
Report No. ____________________

Subcontract No.	Subcontractor	Insurance Exp. Date	Comments

Distribution: ____________________

Action: ____________________

Operating & Maintenance Manual Log

(including spare parts and other project deliverables)

Project No. ______
Project Name ______
Project Location ______
Date ______
Page 1 of ______
Compiled By ______
Report No. ______

Submittal No.	Spec. Section No.	Vendor/Subcontractor Details	Date Rec'd	Sent for Review To Date	Date Returned		Copies Sent To/Number & Date				Resubmit Req'd.		Comments
					Due	Rec'd	Owner	E/A	Vend/Cont.	Other	Yes	No	
							☐	☐	☐	☐			
							☐	☐	☐	☐			
							☐	☐	☐	☐			
							☐	☐	☐	☐			
							☐	☐	☐	☐			
							☐	☐	☐	☐			
							☐	☐	☐	☐			
							☐	☐	☐	☐			
							☐	☐	☐	☐			
							☐	☐	☐	☐			
							☐	☐	☐	☐			
							☐	☐	☐	☐			
							☐	☐	☐	☐			
							☐	☐	☐	☐			
							☐	☐	☐	☐			
							☐	☐	☐	☐			
							☐	☐	☐	☐			

Remarks ______ Distribution ______ Action ______

Shop Drawing Log

Project No. ____________
Project Name ____________
Project Location ____________
Date ____________
Page 1 of ____________
Compiled By ____________
Report No. ____________

Submittal No.	Spec. Section No.	Vendor/Subcontractor Details	Date Rec'd	Sent for Review To	Date Returned		Copies Sent To/Number & Date				Resubmit Req'd.		Comments
				Date	Due	Rec'd	Owner	E/A	Vend/Cont.	Other	Yes	No	
							☐	☐	☐	☐			
							☐	☐	☐	☐			
							☐	☐	☐	☐			
							☐	☐	☐	☐			
							☐	☐	☐	☐			
							☐	☐	☐	☐			
							☐	☐	☐	☐			
							☐	☐	☐	☐			
							☐	☐	☐	☐			
							☐	☐	☐	☐			
							☐	☐	☐	☐			
							☐	☐	☐	☐			
							☐	☐	☐	☐			
							☐	☐	☐	☐			
							☐	☐	☐	☐			
							☐	☐	☐	☐			
							☐	☐	☐	☐			

Remarks ____________

Distribution ____________

Action ____________

Project Accident Report

Project No. ____________
Project Name ____________
Project Location ____________
Date ____________
Page 1 of ____________
Compiled By ____________

Date	Contractor(s) Involved	Accident Description & Name of Individuals Involved in Accident	Number of Individuals Injured	Type of Injury	Recordable Yes/No	Productive Manhours Lost	Cost of Accident	Comments
					☐ Yes ☐ No			
					☐ Yes ☐ No			
					☐ Yes ☐ No			
					☐ Yes ☐ No			
					☐ Yes ☐ No			
					☐ Yes ☐ No			
					☐ Yes ☐ No			
					☐ Yes ☐ No			
					☐ Yes ☐ No			
					☐ Yes ☐ No			
					☐ Yes ☐ No			

Remarks ____________

Distribution ____________

Action ____________

Minutes of Meeting Report

Project No. ______________________
Project Name______________________
Project Location ______________________
Date______________________
Page 1 of ______________________
Compiled by: ______________________
Report No. ______________________

Meeting Subject ____

____

Meeting Location ______________________ Date__________ Time __________

Individuals At Meeting	Title	Representing	Full Time:	Part Time:
			☐	☐
			☐	☐
			☐	☐
			☐	☐
			☐	☐
			☐	☐

Item No.	Minutes/Description	Action By	Action Required Date

Remarks: ____

Distribution: ______________________

Fabrication/Inspection Status Report/ Valves

Vendor/Supplier Name ______________________
____
P.O. No ______________________
Sub-Vendor/Supplier Name ______________________
Location ______________________
Contact ______________________

Project No. ______________
Report No. ______________
Date ______________
Page ______ of ______
Contract No. ______________
Inspected By ______________

Legend

C=Complete S=Started F=Fabrication in Progress N=Not Yet Started D=Delivered to Site

Valve No.	Body Castings	Internals	Operators	Machining	Assembly	Testing	Delivered to Site	Remarks/Comments

Fabrication/Inspection Status Report/ Structural Steel

Vendor/Supplier Name ______________________
____
P.O. No. ______________________
Sub-Vendor/Supplier Name ______________________
Location ______________________
Contact ______________________

Project No. ______________
Report No. ______________
Date ______________
Page ________ of ________
Contract No. ______________
Inspected by ______________

Legend

C=Complete S=Started F=Fabrication in Progress N=Not Yet Started D=Delivered to Site

Description	Shop Drawings Approved for Fabrication	Materials Delivered to Site	Fabrication Work	Welding Work	Priming & Painting	Anchor Bolts	Delivered to Site	Remarks/Comments

Initial Contact Report

Project No. ____________________
Project Name____________________
Project Location ____________________
Date____________________
Compiled by: ____________________
Page_____ of____________________

Fabricator's Name ____________________
Address ____________________
Purchase Order No. ____________________ Revision No. ____________________
Revision Date ____________________ Promised Delivery Date ____________________

Has vendor or supplier received copy of purchase order?

☐ Yes ☐ No ☐ Not applicable

If no, why not? ____________________

What is fabricator's or manufacturer's corresponding shop order number?____________________

Has a detailed engineering schedule been prepared?

☐ Yes ☐ No ☐ Not applicable

If no, why not? ____________________

Has a detailed manufacture, fabrication, or projection schedule been prepared?

☐ Yes ☐ No ☐ Not applicable

If no, why not? ____________________

Does quoted or promised delivery date remain in effect?

☐ Yes ☐ No ☐ Not applicable

If no, why not? ____________________

What is new date if quoted delivery date has changed? ____________________

Does vendor or supplier understand all the submittal data requirements?

☐ Yes ☐ No ☐ Not applicable

If no, why not? ____________________

Location of the manufacturing/fabrication facility ____________________

Contact Name and Phone No: ____________________

What are the addresses and contact names of any sub-vendors or sub-suppliers?

____

Initial Contact Report (cont'd.)

Project No. __________
Project Name__________
Project Location __________
Date__________
Compiled by: __________
Page_____ of__________

Does the vendor or supplier know the delivery address of the finished product?

☐ Yes ☐ No ☐ Not applicable

If no, why not? __________

Does the vendor or supplier understand the invoicing procedure?

☐ Yes ☐ No ☐ Not applicable

If no, why not? __________

Contact names, telephone numbers, and FAX numbers of key personnel in the manufacturing/fabrication facility:

Operations Manager __________	Tel. No. __________	FAX No. __________
Engineering Manager __________	Tel. No. __________	FAX No. __________
Production Supervisor __________	Tel. No. __________	FAX No. __________
QA/QC Manager __________	Tel. No. __________	FAX No. __________
Purchasing Manager __________	Tel. No. __________	FAX No. __________
Transport Manager __________	Tel. No. __________	FAX No. __________

Is manufacturer/fabricator issuing biweekly or monthly status reports?

☐ Yes ☐ No ☐ Not applicable

If no, why not? __________

Comments: __________

Distribution: __________

Action: __________

Expediting Record/Report

Project No. ____________________
Project Name ____________________
Project Location ____________________
Date ____________________
Page _____ of ____________________
Prepared By: ____________________
Report No. ____________________

P.O./S.C. Delivery Date ____________________
Revised Delivery Date ____________________
P.O./S.C. No. ____________________
Equipment No. ____________________
Vendor/Supplier's Name ____________________
Address ____________________
Telephone ____________________
FAX No. ____________________
Contact Name ____________________
Title ____________________

Expediter's Name ____________________
Contact Date ____________________
Contact Time ____________________

Ref.	Comments	Action By

Comments: ____________________

Distribution: ____________________

Action: ____________________

Transport Record

Project No. ______________________
Project Name______________________
Project Location ______________________
Date______________________
Page_____ of ______________________
Prepared by: ______________________
Report No. ______________________

Purchase Order/Contract No. ______________________
Supplier/Vendor ______________________ Record No. ________ of________
Contact ______________________
Telephone No. ______________________
Address ______________________

____

Amount or Value of P.O./Contract $ ______________________
Transport Movement/Shipping Record ______________________

Shipment/ Ref. No.	Material/ Equipment	Gross Weight	Transport/Air Ocean Company	Point of Origin Port	Bill of Lading No.	Freight Cost

Comments:______________________

____

____

____

Distribution: ______________________ Action: ______________________

Subcontract Bid Telephone Confirmation

Project No. ________________
Project Name________________
Project Location ________________
Date________________
Page_____ of__________
Prepared by: ________________
Report No. ________________

Date: ____________ Revision: ____________ Time: ____________

Owner ________________
Subcontractor Name ________________
Address ________________
Telephone No. ________________
Contact ________________
Title ________________
Fax No. ________________
Bid Reference No.________________

Scope of Work Bid: ________________

	Yes	No
All Taxes in Bid	☐	☐
Freight	☐	☐
Plan & Spec	☐	☐
MBE/WBE Contractor	☐	☐
Shop Drawings	☐	☐
Subcontractor's Own General Conditions	☐	☐

Addenda No's. included in bid ________________
Value: $ ________________

Exclusions and omissions to bid: ________________

Bid Value: $ ________________
Alternatives: A. ________________

$ ________________
B. ________________

$ ________________
Information received by ________________(Date/Time) Information.
From: ________________

Note: Subcontractor must confirm bid price within 24 hours.

Comments: ________________

Bid Tabulation Summary

Project No. ______
Project Name ______
Project Location ______
Client ______
Date ______ Rev. ______
Page ______ of ______
Prepared By ______

Description/ Listing of Items	Bidder ______ Address ______ Bid No. ______	Bidder ______ Address ______ Bid No. ______	Bidder ______ Address ______ Bid No. ______	Bidder ______ Address ______ Bid No. ______	Bidder ______ Address ______ Bid No. ______
Base Bid:					
Freight Costs:					
Packing Costs:					
Taxes:					
Escalation:					
Bonds:					
Other:					
Other:					
Other:					
Total Cost					
Meets Spec:	☐ Yes ☐ No	☐ Yes ☐ No	☐ Yes ☐ No	☐ Yes ☐ No	☐ Yes ☐ No
Bid Expir. Date:					
F.O.B. Point:					
Warranty Period:					
Other:					

Comments: ______

Approvals/Recommendations

Award to Bid No. ______
Bid Value: $ ______
Signed by Purchasing ______ Date ______
Signed by Engineering ______ Date ______
Project Manager ______ Date ______

Subcontractor Questionnaire Data Form

Project No. ____
Project Name ____
Project Location ____
Date ____
Prepared By ____
Page ____ of ____

Organization

Name of Company ____
Street Address ____
City and State ____ Zip Code ____
Telephone No. ____ FAX No. ____

1. Indicate type of business organization.

   ☐ **Corporation**. List names of officers.

   President ____ Secretary ____
   General Manager ____ Treasurer ____
   Place of Incorporation ____ Date ____

   ☐ **Partnership**. List names of partners.

   ____
   ____
   ____
   ____

   ☐ **Sole Owner**. Name ____

2. Names of owners (stockholders holding over 10% of stock):

   ____ ____
   ____ ____
   ____ ____

3. Principals of company (officers' names, titles, qualifications, experience and years):

   ____
   ____
   ____
   ____

4. Subsidiaries (indicate whether wholly owned or percent controlled):

   ____
   ____

5. Number of years in business under your present name: ____ years.
6. The company is ____% minority owned.
7. Indicate the number of permanent employees currently on payroll:

   Management ____
   Engineers ____
   Draftsmen ____
   Office admin. staff ____
   Field supervisors ____
   Field labor force ____
   Total ____

Subcontractor Questionnaire Data Form (cont'd)

Project No. ______________________
Project Name______________________
Project Location ______________________
Date______________________
Prepared By______________________
Page_____ of______________________

CLASSIFICATION OR TYPE OF WORK PERFORMED

1. Check type of construction work your company employees perform.

- ☐ Architectural work
- ☐ Carpentry work
- ☐ Concrete work
- ☐ Conveyors/Elevators
- ☐ Demolition/Relocations
- ☐ Electrical work
- ☐ Excavation work
- ☐ Fencing work
- ☐ Fire protection work
- ☐ Glass/glazing work
- ☐ Hazardous Waste Removal
- ☐ H.V.A.C. work
- ☐ Inspection & testing work
- ☐ Insulation/heat tracing work
- ☐ Lab equipment installation work
- ☐ Landscaping work
- ☐ Masonry/brickwork
- ☐ Mechanical work
- ☐ Millwright work
- ☐ Painting work
- ☐ Paving work
- ☐ Pile driving work
- ☐ Piping systems
- ☐ Plumbing work
- ☐ Roofing work
- ☐ Sheet metal work
- ☐ Site work
- ☐ Sprinkler work
- ☐ Structural steel work
- ☐ Tunnelling work
- ☐ Others (List)

____
____

2. Percent of work performed as a general contractor ____________%
3. Percent of work performed as a subcontractor ____________%
4. List type of work usually subcontracted to others ______________________

____
____

WORK HISTORY

1. List the important projects completed by your organization within the last five years including the contract value.

Name of Client	Person to Contact	Project Title and Scope of Work	Contract Value	Year Work Performed

FINANCIAL

1. Submit last three annual financial reports and current profit and loss statement (audited report preferred).
2. a. What is the maximum dollar value of a project you believe your company is capable of handling?
   $______________________
   b. Over what period of time?______________________
3. Average annual dollar volume of work for the past five years $______________________
4. Is there any litigation now in progress or pending with clients, subcontractors, or suppliers? ☐ Yes ☐ No
   If yes, give details. ______________________

____
____

Subcontractor Questionnaire Data Form (cont'd)

Project No. ______________________
Project Name______________________
Project Location ______________________
Date______________________
Prepared By______________________
Page_____ of______________________

5. a. Do you have an established bonding company? ☐ Yes ☐ No
   b. If yes, name of bonding company ______________________
   Address ______________________
   Contact ______________________ Phone No. ____________ Bonding Capacity $ ____________
6. Indicate banking references ______________________

____

____

CONSTRUCTION EQUIPMENT AND MACHINERY

1. List owned construction equipment with capacity, age, type and attachments.

____

____

LABOR AND UNION AFFILIATIONS

1. Does your organization perform work as an open shop? ☐ Yes ☐ No Closed shop? ☐ Yes ☐ No
   If both, is work performed under same name? ☐ Yes ☐ No If different names, please list both.

____

____

2. Do you have any union national agreements? ☐ Yes ☐ No If yes, with which crafts?

____

____

3. If you are signator on local agreements, indicate the following:

Craft and Local	Holder of your bargaining rights

SAFETY AND INSURANCE INFORMATION

1. Person responsible for safety program ______________________
   Title ______________________ Phone No. ______________________
2. Person to be contacted for matters involving insurance______________________
   Phone No. ______________________ FAX No. ______________________
3. Insurance agent's name and address ______________________

____

	Last Year	Previous 5 Yrs.
4. a. Number of lost workday cases (injuries involving days away from work)		
b. Number of cases with medical treatment only		
c. Number of fatalities		

This statement was completed by:

Name______________________ Title ______________________

Signature______________________ Date ______________________

Contracting Implementation Plan

Project No. ____________
Project Name ____________
Revision No. ____________
Date ____________

Evaluation of Owner's Needs					Bidding Phase					Award Phase		Construction Implementation Phase											
Months					Weeks					Weeks		Months (varies with magnitude of project)											
1	2	3	4	5	1	2	3	4	5	1	2	1	2	3	4	5	6	7	8	9	10	11	12

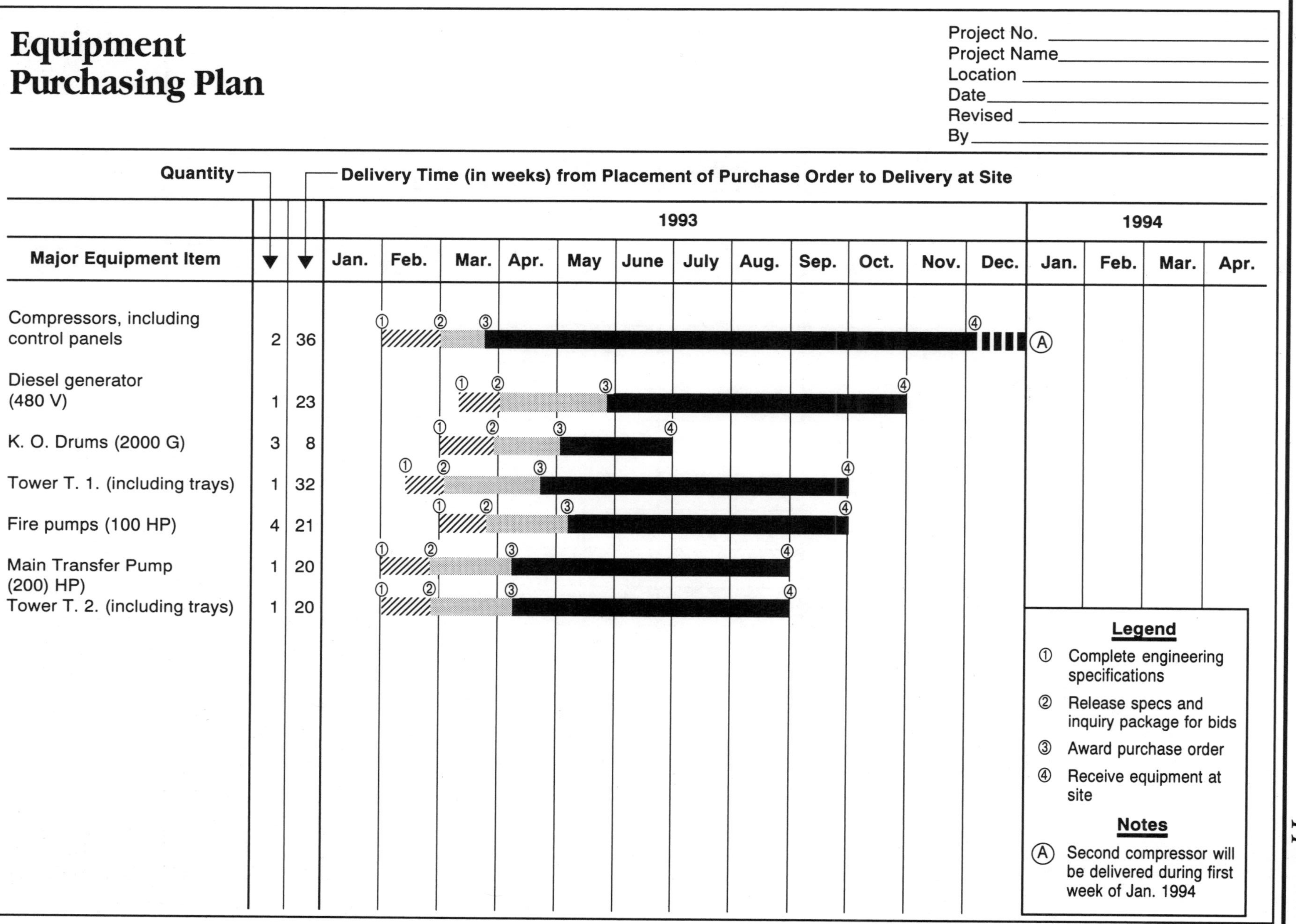
Equipment Purchasing Plan
Project No.
Project Name
Location
Date
Revised
By
Quantity
Delivery Time (in weeks) from Placement of Purchase Order to Delivery at Site
1993
1994
Major Equipment Item
Jan.
Feb.
Mar.
Apr.
May
June
July
Aug.
Sep.
Oct.
Nov.
Dec.
Jan.
Feb.
Mar.
Apr.
Compressors, including control panels
2
36
Diesel generator (480 V)
1
23
K. O. Drums (2000 G)
3
8
Tower T. 1. (including trays)
1
32
Fire pumps (100 HP)
4
21
Main Transfer Pump (200) HP)
1
20
Tower T. 2. (including trays)
1
20
Legend
① Complete engineering specifications
② Release specs and inquiry package for bids
③ Award purchase order
④ Receive equipment at site
Notes
Ⓐ Second compressor will be delivered during first week of Jan. 1994

Responsibility Matrix for Project Procurement Activities

Project No. ____________
Project Name ____________
Revision No. ____________
Date ____________

Construction Project Equipment/Materials	Owner-Furnished Equipment		Managing Contractor-Furnished Items		Admin. Bldg. Contractor-Furnished Items		Pipeline Contractor-Furnished Items		Pump Station Contractor-Furnished Items		Railroad Contractor-Furnished Items		Comments
	H.O. Purch'd	Field Purch'd	H.O. Purch'd	Field Purch'd	H.O. Purch'd	Field Purch'd	H.O. Purch'd	Field Purch'd	H.O. Purch'd	Field Purch'd	H.O. Purch'd	Field Purch'd	
Dryers													
Grinders													
Towers													
Vessels													
Compressors													
Pumps													
Heat Exchangers													
Storage Tanks													
Concrete													
Steel Piping													
Main Pipeline Pipe													
Building Materials • Lumber • Paving/Stone • Blockwork/Brick													
Elevators													
Structural Steel													
Pipe (less than 12")													
Valves													
Fittings													
Supports													
Main Elec. Equipt.													
Instrumentation													
Insulation													
Paint													
Temp. Main Camp													
Temp. Pipeline Camps													
Temp. Pump/Stat. Camp													
Laboratory Equipt.													
Railroad Equipt.													
Office Equipt.													
Temp. Office Equipt.													
Plant Truck													
Plant Autos													
Spare Parts													
Plant Start-up Materials													

(Company Letterhead)

To: ______________________________ Project No. ______________

______________________________ Project Name______________

______________________________ Bid Package No. ______________

Subject: ______________________________ Contract No. ______________

______________________________ Project Location ______________

______________________________ Page 1 of ______________

Dear Sir or Madam:

This letter is to inform you that it is our intent to award the above referenced contract to your company. This letter of intent is based on our Bid Package Number referenced above and subsequent clarifications as referenced here:

____

____

____

This letter of intent is contingent upon a contract/subcontract being consummated and agreed to by the two parties concerned.

The contract/subcontract is being prepared and will be forwarded to you on or before ______________. The contract/subcontract will include any modifications and agreed adjustments that were made during our post-bid meetings or discussions. Prior to your commencement of the contract/subcontract, we will require satisfactory proof of insurance, together with the following:

____

____

____

When you receive our contract/subcontract, please sign and return to this office: Attention______________________.

If you have any comments or questions, please contact us as soon as possible.

Sincerely,

____

Distribution: ______________________________ Action: ______________________________

______________________________ ______________________________

(Company Letterhead)

To: ______________________________

____

____

Subject: ______________________________

____

____

Project No. ______________________________
Project Name______________________________
Contract No. ______________________________
Purchase Order No. ______________________________
Project Location ______________________________
Page 1 of ______________________________

Dear Sir or Madam:

The subcontract for the contract number referenced above has been awarded to another company. We would like to take this opportunity to thank you for your proposal. In the future, we would not hesitate to obtain future proposals from your company if the need arises.

Sincerely,

____

Distribution: ______________________________

____

Action: ______________________________

____

(Company Letterhead)

To: ______________________________

____

____

Subject: ______________________________

____

____

Project No. ______________

Project Name ______________

Contract No. ______________

Purchase Order No. ______________

Project Location ______________

Page 1 of ______________

Dear Sir or Madam:

In accordance with the terms and conditions of the above referenced contract/purchase order, please forward the following:

☐ Submittals and Samples: ______________________________

____

____

☐ Operating Manuals: ______________________________

____

____

☐ As-Built Drawings: ______________________________

____

____

☐ Other: ______________________________

____

____

This information is required no later than ______________.

If you have questions or comments, please contact us at your earliest convenience.

Sincerely,

____

Distribution: ______________________________	Action: ______________________________
______________________________	______________________________

(Company Letterhead)

To: ______________________________ Project No. ______________
______________________________ Project Name______________
______________________________ Contract No. ______________
______________________________ Purchase Order No. ______________
Subject: ______________________________ Project Location ______________
______________________________ Page 1 of ______________
____

Dear Sir or Madam:

We have received your Application for Payment No. ___________.

This application, however, has been adjusted and modified for the following reasons:

____

____

____

____

The adjusted payment that is due to your company is $_________________. An adjusted and marked-up copy of your Application for Payment is enclosed for your records.

Sincerely,

____

Remarks: ______________________________

Distribution: ______________________________ Action: ______________________________

(Company Letterhead)

To: ______________________________ Project No. ______________

______________________________ Project Name ______________

______________________________ Contract No. ______________

Purchase Order No. ______________

Subject: ______________________________ Project Location ______________

______________________________ Page 1 of ______________

____

Dear Sir or Madam:

Our records indicate that your insurance policy related to the above contract will expire on ____________.

Please notify your insurance company and request that an updated Certificate of Insurance be forwarded to the above address, to the attention of ______________________ as soon as possible.

Sincerely,

____

Remarks: ______________________________

____

Distribution: ______________________ Action: ______________________

______________________ ______________________

Appendix B

Professional Associations

ACRI Air-Conditioning and Refrigeration Institute
1501 Wilson Blvd., 6th Floor
Arlington, VA 22209

AMCA Air Movement and Control Association
30 West University Drive
Arlington Heights, IL 60004

AA Aluminum Association
900 19th Street, NW, Suite 300
Washington, DC 20006

AACE American Association of Cost Engineers
209 Prairie Ave., Suite 100
Morgantown, WV 26505

AAN American Association of Nurserymen, Inc.
1250 I St., NW, Suite 500
Washington, DC 20005

ACI American Concrete Institute
Box 19150
Redford Station
Detroit, MI 48219

ACEC American Consulting Engineers Council
1015 15th Street, NW
Washington, DC 20005

AIA American Institute of Architects
1735 New York Avenue, NW
Washington, DC 20006

AIC	American Institute of Constructors 9887 N. Gandy, Suite 104 St. Petersburg, FL 33702
AISC	American Institute of Steel Construction 1 E. Wacker Dr., Suite 3100 Chicago, IL 60601
AITC	American Institute of Timber Construction 11818 SE Mill Plaine Blvd., Suite 415 Vancouver, WA 98684
AISI	American Iron and Steel Institute 1133 15th Street, NW Washington, DC 20005
ANSI	American National Standards Institute 11 West 42nd St. New York, NY 10036
APA	American Plywood Association Box 11700 Tacoma, WA 98411
ASCE	American Society of Civil Engineers 345 East 47th Street New York, NY 10017
ASHRAE	American Society of Heating Refrigerating and Air Conditioning Engineers 1791 Tullie Circle, NE Atlanta, GA 30329
ASLA	American Society of Landscape Architects 4401 Connecticut Ave., NW Washington, DC 20008
ASME	American Society of Mechanical Engineers 345 East 47th Street New York, NY 10017
ASPE	American Society of Professional Estimators 11141 Georgia Ave, Suite 412 Wheaton, MD 20902

ASTM	American Society for Testing and Materials 1916 Race Street Philadelphia, PA 19103
AWWA	American Water Works Association 6666 West Quincy Avenue Denver, CO 80235
AWS	American Welding Society 550 Lejeune Rd, NW Miami, FL 33126
AWC	American Wood Council 1250 Connecticut Avenue, NW, Suite 300 Washington, DC 20036
AWPA	American Wood Preservers' Association P.O. Box 286 Woodstock, MD 21163
AWI	Architectural Woodwork Institute P.O. Box 1550 Centerville, VA 22020
AI	Asphalt Institute Research Park Dr. P.O. Box 14052 Lexington, KY 40512
ABC	Associated Builders and Contractors, Inc. 729 15th Street, NW Washington, DC 20005
AED	Associated Equipment Distributors 615 W. 22nd Street Oak Brook, IL 60521
AGC	Associated General Contractors of America 1957 E Street, NW Washington, DC 20006
ALCA	Associated Landscape Contractors of America 405 N. Washington Steet Falls Church, VA 22046

BIA	Brick Institute of America 11490 Commerce Park Dr., Suite 300 Reston, VA 22091
BHMA	Builders Hardware Manufacturers Assoc. 355 Lexington Ave., 17th Floor New York, NY 10017
CFMA	Construction Financial Management Association 40 Brunswick Ave., Suite 202 Edison, NJ 08818
CIMA	Construction Industry Manufacturers Assoc. 111 E. Wisconsin Ave., Suite 940 Milwaukee, WI 53202
CMAA	Construction Management Association of America 1893 Preston White, Suite 130 Reston, VA 22091
CSI	Construction Specifications Institute 601 Madison Street Alexandria, VA 22314
CRSI	Concrete Reinforcing Steel Institute 933 Plum Grove Road Schaumberg, IL 60173
CDA	Copper Development Association, Inc. P.O. Box 1840 Greenwich, CT 06836
DHI	Door & Hardware Institute 14170 Newbrook Drive Chantilly, VA 22021
FMERO	Factory Mutual Engineering & Research Group 1151 Boston-Providence Turnpike Norwood, MA 02062
FGMA	Flat Glass Marketing Association White Lakes Professional Building 3310 SW Harrison Street Topeka, KS 66611

GSA	General Service Administration F Street and 18th, NW Washington, DC 20405
GA	Gypsum Association 810 1st Street, NE, Suite 510 Washington, DC 20002
IEEE	Institute of Electrical and Electronic Engineers 345 East 47th Street New York, NY 10017
IMI	International Masonry Insitute 823 15th Street, NW Washington, DC 20005
MCAA	Mechanical Contractors Association of America, Inc. 1385 Piccard Drive Rockville, MD 20832
MLSFA	Metal Lath/Steel Framing Association 600 South Federal, Suite 400 Chicago, IL 60605
NAAMM	National Association of Architectural Metal Manufacturers 600 South Federal, Suite 400 Chicago, IL 60605
NAHB	National Association of Home Builders 15th & M Street, NW Washington, DC 20005
NAPHCC	National Association of Plumbing-Heating- Cooling Contractors P.O. Box 6808 Falls Church, VA 22046
NECA	National Electrical Contractors Association 7315 Wisconsin Avenue Bethesda, MD 20814

NEMA	National Electrical Manufacturers Association 2101 L Street, NW, Washington, DC 20037
NIBS	National Institute of Building Sciences 1201 L Street NW, Suite 400 Washington, DC 20005
NFPA	National Fire Protection Association One Batterymarch Park Quincy, MA 02269
NFPA	National Forest Products Association 1250 Connecticut Avenue, NW Washington, DC 20036
NPCA	National Precast Concrete Association 825 E. 64th Street Indianapolis, IN 46220
NRCA	National Roofing Contractors Association 10255 West Higgins Road, Suite 600 Rosemont, IL 60018
NSPE	National Society of Professional Engineers 1420 King Street Alexandria, VA 22314
NSWMA	National Solid Wastes Management Association 1730 Rhode Island Avenue, NW Washington, DC 20036
NUCA	National Utility Contractors Association 1235 Jefferson Davis Hwy., Suite 606 Arlington, VA 22202
PCA	Portland Cement Association 5420 Old Orchard Road Skokie, IL 60077
PCEA	Professional Construction Estimators Association P.O. Box 1107 Cornelius, NC 28031

PS	Product Standard U.S. Department of Commerce 14th & Constitutional Avenue Washington, DC 20230
SIGMA	Sealed Insulating Glass Manufacturers Association 401 N. Michigan Ave. Chicago, IL 60601
SMACNA	Sheet Metal and Air Conditioning Contractors National Association, Inc. 4201 Lafayette Center Drive Chantilly, VA 22021
SDI	Steel Door Institute 30200 Detroit Road Cleveland, OH 44145
SSPC	Steel Structures Painting Council 4400 Fifth Avenue Pittsburgh, PA 15213
TCA	Tile Council of America, Inc. P.O. Box 326 Princeton, NJ 08542-0326
UL	Underwriters Laboratories, Inc. 333 Pfingsten Road Northbrook, IL 60062
USFS	U.S. Forest Service Forest Products Laboratory P.O. Box 5130 Madison, WI 53705
WCLIB	West Coast Lumber Inspection Bureau 6980 SW Varns Street Box 23145 Portland, OR 97223
WWPA	Western Wood Products Association Yeon Bldg. 522 SW 5th Ave. Portland, OR 97204

Appendix C

Information for Overseas Construction

U.S. Contacts

U.S. Department of Commerce

The scope of information and services provided by the Department is very comprehensive. The Department serves and promotes U.S. organizations, firms and individuals in exporting and trading with other nations. The information and programs that will be of interest to construction-related organizations is concentrated in the Department's International Trade Administration (ITA) of which the subdivision called the U.S. and Foreign Commercial Service (U.S. & FCS) maintains a database and network of international trade experts both in the U.S. and overseas. By making contact with one of the 60-plus Department of Commerce district offices a construction-related organization can gain useful information about any foreign country. The address of the U.S. Department of Commerce is 14th and Constitution Ave. N.W., Washington, DC 20230. Tel. (202) 277-5494.

U.S. and Foreign Commercial Service

Capital and International Construction. Tel. (202) 377-5023

International Major Projects

Tel. (202) 377-5225

African Near East and South Asia

Tel. (202) 377-4836

East Asia and Pacific

Tel. (202) 377-8422

Europe

Tel. (202) 377-1599

Western Hemisphere

Tel. (202) 377-2736

Export-Import Bank of the United States

The Export-Import Bank (Eximbank) is the U.S. Government agency that facilitates the financing of exports of U.S. services, products, and goods. Eximbank basically assists U.S. companies to compete with their foreign competitors by providing loans and certain kinds of insurance. The address of Eximbank is: Export-Import Bank of the United States, 811 Vermont Ave. N.W., Washington, DC 20571. Tel. (202) 566-8860.

U.S. Small Business Administration

The U.S. Small Business Administration (SBA) helps small, midsize, and minority organizations overcome the complexities of exporting and financing export sales. The two major programs that the SBA provides to small, midsize, or minority organizations are (1) business development assistance and (2) financial assistance. The address of the U.S. Small Business Administration is:

U.S. Small Business Administration Office of International Trade
409 Third Street S.W., 6th Floor
Washington, DC 20416
Tel. (202) 205-6720

U.S. Agency for International Aid

The U.S. Agency for International Aid (AID) administers the U.S. foreign economic assistance program to the less developed countries in Africa, Latin America, Asia, the Near East, the Caribbean Basin, and the emerging Eastern European Democracies. The address of the U.S. Agency for International Aid is:

U.S. Agency for International Aid
Department of State Building
320 21st Street N.W.
Washington, DC 20523
Tel. (202) 235-1840

Overseas Private Investment Corporation

Through the Overseas Private Investment Corporation (O.P.I.C.), the U.S. Government encourages and assists U.S. private investments in some of the less developed nations. The address of the Overseas Private Investment Corporation is:

Overseas Private Investment Corporation
1129 20th Street N.W., 7th Floor
Washington, DC 20527
Tel. (202) 457-7010

Foreign Embassy Contacts

Algeria

American Embassy, Commercial Section
4 Chemin Cheikh Bachir El Ibrahimi
Algiers, Algeria
Tel. 213 260 1863

Argentina

American Embassy, Commercial Section
4300 Columbia, 1425
Buenos Aires, Argentina
Tel. 54-1-773-1063

Embassy of Argentina, Commercial Section
1667 K Street, N.W.
Suite 610
Washington, DC 20006
Tel. (202) 939-6400

Australia

American Consulate General – Sydney, Commercial Section
36th Floor, T&G Tower
Hyde Park Square
Sydney 2000
N.S.W., Australia
Tel. 61-2-261-9200

Embassy of Australia, Commercial Section
1601 Massachusetts Avenue, N.W.
Washington, DC 20036
Tel. (202) 797-3201

Austria

American Embassy, Commercial Section
Boltzmanngasse 16
A-1091
Vienna, Austria
Tel. 43-222-31-55-11

Barbados

American Embassy, Commercial Section
Broad Street
Bridgetown, Barbados
Tel. 809-436-4950

Belgium

American Embassy, Commercial Section
27 Boulevard du Regent
B-1000 Brussels, Belgium
Tel. 32-2-513-3830

Embassy of Belgium, Commercial Section
3330 Garfield Street, N.W.
Washington, DC 20008
Tel. (202) 333-6900

Brazil

American Embassy, Commercial Section
Avenida das Nacoes, Lote 3
Brasilia, Brazil
Tel. 55-61-321-7272

Cameroon

American Embassy, Commercial Section
Rue Nachtigal
Yaounde, Cameroon
Tel. 237-23-40-14

Canada

American Embassy, Commercial Section
100 Wellington Street
Ottawa, Ontario
Canada K1PT1
Tel. (613) 238-5335

Embassy of Canada, Commercial Section
1746 Massachusetts Avenue, N.W.
Washington, DC 20036
Tel. (202) 785-1400

Chile

American Embassy, Commercial Section
Edificio Codina
Agustinas 1343
Santiago, Chile
Tel. 56-2671-0133

Embassy of Chile, Commercial Section
1732 Massachusetts Avenue, N.W.
Washington, DC 20036
Tel. (202) 785-1746

China, People's Republic of

American Embassy, Commercial Section
Guang Hua Lu 17
Beijing, China
Tel. 86-1-532-381 x490

Embassy of the People's Republic of China, Commercial Section
2300 Connecticut Avenue, N.W.
Washington, DC 20008
Tel. (202) 328-2520

Colombia

American Embassy, Commercial Section
Calla 38
No. 8-61
Bogota, Colombia
Tel. 57-1-232-6550

Embassy of Colombia, Commercial Section
2118 Leroy Place, N.W.
Washington, DC 20008
Tel. (202) 387-8338

Costa Rica

American Embassy, Commercial Section
Pavas
San Jose, Costa Rica
Tel. 506-20-3939

Cote d'Ivoire

American Embassy, Commercial Section
5 Rue Jesse Owens
Abidjan
Cote d'Ivoire
Tel. 225-21-46-16

Czechoslovakia
American Embassy, Commercial Section
Trziste 15
12548 Prague
Czechoslovakia
Tel. 42-2-53-6641

Denmark
American Embassy, Commercial Section
Dag Hammarskjold Alie 24
2100 Copenhagen, Denmark
Tel. 45-31-42-31-44

Embassy of Denmark, Commercial Section
3200 Whitehaven Street, N.W.
Washington, DC 2008
Tel. (202) 234-4300

Dominican Republic
American Embassy, Commercial Section
Calle Cesar Nicolas Penson con Calle Leopoldo Navarro
Santo Domingo, Dominican Republic
Tel. (809) 541-2171

Embassy of the Dominican Republic, Commercial Section
1715 22nd Street, N.W.
Washington, DC 20007
Tel. (202) 332-6280

Equador
American Embassy Ecuador
120 Avenida Patria
Quito, Ecuador
Tel. 593-2-561-404

Embassy of Ecuador, Commercial Section
2535 15th Street, N.W.
Washington, DC 20009
Tel. (202) 234-7200

Egypt
American Embassy, Commercial Section
5 Sharia Latin America
Cairo, Republic of Eqypt
Tel. 20-2-354-1583

Embassy of Egypt, Commercial Section
2715 Connecticut Avenue, N.W.
Washington, DC 20008
Tel. (202) 265-9111

Finland
American Embassy, Commercial Section
Itained Puistotie 14A
SF-00140 Helsinki, Finland
Tel. 358-0-171-821

France

American Embassy, Commercial Section
2 Avenue Gabriel
75382 Paris Cedex 08
France
Tel. 33-1-42-96-1202

Embassy of France, Commercial Section
4101 Reservoir Road, N.W.
Washington, DC 20007
Tel. (202) 944-6000

Germany

American Embassy, Commercial Section
Delchmannsaue
5300 Bonn 2
Germany
Tel. 49-228-339-2895

Embassy of the Federal Republic of Germany
46-45 Reservoir Road, N.W.
Washington, DC 20007
Tel. (202) 298-4000

Greece

American Embassy, Commercial Section
91 Vasillis Sophia Boulevard
10160 Athens, Greece
Tel 30-1-723-9705

Guatemala

American Embassy, Commercial Section
7-01 Avenida de la Reforma, Zone 10
Guatemala City, Guatemala
Tel. 502-2-34-84-79

Honduras

American Embassy, Commercial Section
Avenida La Paz
Tegucigalpa, Honduras
Tel. 504-32-3120

Hong Kong

American Consulate General, Commercial Section
26 Garden Road
Hong Kong
Tel. 852-521-1467

Hungary

American Embassy, Commercial Section
Bajza Utca 31
H-1062 Budapest, Hungary
Tel. 36-1-122-8600

India

American Embassy, Commercial Section
Shanti Path,
Chanahyapuri
110021 New Delhi, India
Tel. 91-11-600-651

Embassy of India, Commercial Section
2107 Massachusetts Avenue, N.W.
Washington, DC 20008
Tel. (202) 939-7000

Indonesia

American Embassy, Commercial Section
Medan Merdeka Selatan 5
Jakarta, Indonesia
Tel. 62-21-360-360

Embassy of Indonesia, Commercial Section
2020 Massachusetts Avenue, N.W.
Washington, DC
Tel. (202) 775-5200

Ireland

American Embassy, Commercial Section
42 Elgin Road
Ballsbridge
Dublin, Ireland
Tel. 353-1-687-122

Embassy of Ireland, Commercial Section
2234 Massachusetts Avenue, N.W.
Washington, DC 20008
Tel. (202) 462-3939

Israel

American Embassy, Commercial Section
71 Hayarkon Street
Tel Aviv, Israel
Tel. 972-3-654-338

Embassy of Israel
3514 International Drive, N.W.
Washington, DC 20008
Tel. (202) 364-5500

Italy

American Embassy, Commercial Section
Via Veneto 119/A
00187 Rome, Italy
Tel. 39-6-4674-2202

Embassy of Italy Commercial Section
1601 Fuller Street, N.W.
Washington, DC 20009
Tel (202) 328-5505

Jamaica
American Embassy, Commercial Section
2 Oxford Road, 3rd Floor
Kingston, Jamaica
Tel. (809) 929-4850

Japan
American Embassy, Commercial Section
10-5 Akasaka, 1-Chrome
Minato-ku (107)
Tokyo, Japan
Tel. 81-3-3224-5050

Embassy of Japan, Commercial Section
2520 Massachusetts Avenue, N.W.
Washington, DC 20008
Tel. (202) 939-6700

Kenya
American Embassy, Commercial Section
Moi Haile Selassie Avenue
Nairobi, Kenya
Tel. 254-2-334-141

Korea
American Embassy, Commercial Section
82 Sejong-Ro
Chongro-ku
Seoul, Korea
Tel. 82-2-732-2601

Kuwait
American Embassy, Commercial Section
P.O. Box 77 SAFAT
Kuwait
Tel. 965-242-4151

Embassy of Kuwait, Commercial Section
2940 Tilden Street, N.W.
Washington, DC 20008
Tel. (202) 966-0702

Malaysia
American Embassy, Commercial Section
AIA Building 376 Jalan Tun Razah
P.O. Box 10035
50700 Kuala Lumpur, Malaysia
Tel. 60-248-9011

Embassy of Malaysia, Commercial Section
2401 Massachusetts Avenue, N.W.
Washington, DC 20008
Tel. (202) 328-2700

Mexico

American Embassy, Commercial Section
Paseo de la Reforma 305
Mexico City 06500, Mexico
Tel. 52-5-211-0042

Embassy of Mexico, Commercial Section
1911 Pennsylvania Avenue, N.W.
Washington, DC 20006
Tel. (202) 728-1600

Morocco

American Consulate General – Casablance, Commercial Section
8 Boulevard Moulay Youssef
Casablance, Morocco
Tel. 212-26-45-50

Netherlands

American Embassy, Commercial Section
Lange Voorhout 102
The Hague, The Netherlands
Tel. 31-70-310-9417

Embassy of the Netherlands Commercial Section
4200 Linnean Avenue, N.W.
Washington, DC 20008
Tel. (202) 244-5300

New Zealand

American Consulate General – Auckland, Commercial Section
4th Floor, Yorkshire General Building
Auckland, New Zealand
Tel. 64-9-303-2038

Embassy of New Zealand, Commercial Section
37 Observatory Circle, N.W.
Washington, DC 20008
Tel. (202) 328-4800

Nigeria

American Embassy, Commercial Section
2 Eleke Crescent
P.O. Box 554
Lagos, Nigeria
Tel. 234-1-616-477

Embassy of Nigeria, Commercial Section
2201 M Street, N.W.
Washington, DC 20037
Tel. (202) 822-1500

Norway

American Embassy, Commercial Section
Drammensveien 18
Oslo 2, Norway
Tel. 47-2-44-85-50

Embassy of Norway, Commercial Section
2720 34th Street, N.W.
Washington, DC 20008
Tel. (202) 333-6000

Pakistan

American Consulate General – Karachi,
8 Abdullah Harroon Road
Karachi, Pakistan
Tel. 92-21-518-180

Embassy of Pakistan, Commercial Section
2315 Massachusetts Avenue, N.W.
Washington, DC 20008
Tel. (202) 939-6200

Panama

American Embassy, Commercial Section
Avenida Balboa y Calle 38
Apartado 6959
Panama 5, Republic of Panama
Tel. 507-27-1777

Embassy of Panama, Commercial Section
2862 McGill Terrace, N.W.
Washington, DC 20008 USA
Tel. (202) 483-1407

Peru

American Embassy, Commercial Section
P.O. Box 1995
Lima 100 Peru

Embassy of Peru, Commercial Section
1700 Massachusetts Avenue, N.W.
Washington, DC 20036
Tel. (202) 833-9860

Philippines

American Embassy, Commercial Section
395 Buendia Avenue
Extension Makati
Manila, The Philippines
Tel. 63-2-818-6674

Embassy of the Philippines, Commercial Section
1617 Massachusetts Avenue, N.W.
Washington, DC 20036
Tel. (202) 483-1414

Poland

American Embassy, Commercial Section
Ulica Wiejska 20
Warsaw, Poland
Tel. 48-22-21-45-15

Portugal

American Embassy, Commercial Section
Avenida das Forcas Armadas
1600 Lisbon, Portugal
Tel. 351-1-726-6600

Embassy of Portugal, Commercial Section
1914 Connecticut Avenue, N.W.
Washington, DC 20008
Tel. (202) 328-8610

Romania

American Embassy, Commercial Section
Strada Tudor Arghezi 7-9
Bucharest, Romania
Tel. 40-0-10-40-40

Russia

U.S. Commercial Office – Moscow
Ulitsa Chaykovskogo 15
Moscow, Russia
Tel. 7-096-255-4848

Russian Trade Representative in the U.S.A.
2001 Connecticut Avenue, N.W.
Washington, DC 20008
Tel. (202) 234-8304

Saudi Arabia

American Embassy, Commercial Section
Collector Road M, Riyadh Diplomatic Quarter
Riyadh, Saudi Arabia
Tel. 966-1-488-3800

Embassy of Saudi Arabia, Commercial Section
601 New Hampshire Avenue, N.W.
Washington, DC 20037
Tel. (202) 337-4088

Singapore

American Embassy, Commercial Section
One Colombo Court #05-12
Singapore 0617

Embassy of Singapore, Commercial Section
1824 R Street, N.W.
Washington, DC 20009
Tel. (202) 667-7555

South Africa

American Consulate General – Johannesburg, Commercial Section
Kine Center, 11th Floor
Commissioner and Kruis Streets
P.O. Box 2155
Johannesburg, South Africa
Tel. 27-11-331-3937

Embassy of South Africa, Commercial Section
3051 Massachusetts Avenue, N.W.
Washington, DC 20016
Tel. (202) 232-4400

Spain

American Embassy, Commercial Section
Serrano 75
Madrid, Spain

Embassy of Spain, Commercial Section
2700 15th Street, N.W.
Washington, DC 20009
Tel. (202) 265-8600

Sweden

American Embassy, Commercial Section
Strandvagen 101
Stockholm, Sweden
Tel. 46-8-783-5346

Embassy of Sweden, Commercial Section
600 New Hampshire Avenue, N.W.
Washington, DC 20037
Tel. (202) 944-5600

Switzerland

American Embassy, Commercial Section
Julinaeumstrasse 93
3005 Bern, Switzerland
Tel. 41-31-437-341

Embassy of Switzerland, Commercial Section
2900 Cathedral Avenue, N.W.
Washington, DC 20008
Tel. (202) 745-7900

Taiwan

American Chamber of Commerce in Taiwan
Rm. 1012, Chia Hsin Building Annex
96 Chung Shan Road, Section 2
P.O. Box 17-277
Taipei, Taiwan
Tel. 886-2-551-2515

Coordination Council for North American Affairs
Economic Division
4301 Connecticut Avenue, N.W.
Suite 420
Washington, DC 20008
Tel. (202) 686-6400

Thailand

American Embassy, Commercial Section
95 Wireless Road
Bangkok, Thailand
Tel. 66-2-253-4929

Embassy of Thailand, Commercial Section
2300 Kalorama Road, N.W.
Washington, DC 20008
Tel. (202) 467-6790

Trinidad and Tobago

American Embassy, Commercial Section
15 Queen's Park West
P.O. Box 752
Port-of-Spain, Trinidad and Tobago
Tel. 809-622-6371

Embassy of Trinidad and Tobago, Commercial Section
1708 Massachusetts Avenue, N.W.
Washington, DC 20036
Tel. (202) 467-6490

Turkey

American Embassy, Commercial Section
110 Ataturk Boulevard
Ankara, Turkey
Tel. 90-4-167-0949

Embassy of Turkey, Commercial Section
2523 Massachusetts Avenue, N.W.
Washington, DC 20008
Tel. (202) 483-5366

United Arab Emirates

American Embassy, Commercial Section
Blue Tower Building, 8th Floor
Shaikh Khalifa Bin Zayed Street
Abu Dhabi, U.A.E.
Tel. 971-2-345545

Embassy of the United Arab Emirates, Commercial Section
600 New Hampshire Avenue, N.W., Suite 740
Washington, DC 20037
Tel. (202) 338-6500

United Kingdom

American Embassy, Commercial Section
24/31 Grosvenor Square
London W. 1A 1AE, England
Tel 44-71-499-9000

Embassy of Great Britain, Commercial Section
3100 Massachusetts Ave, N.W.
Washington, DC 20008
Tel. (202) 462-1340

Venezuela

American Embassy, Commercial Section
Avenida Francisco de Mirandola and Avenida Principal de la Floresta
P.O. Box 62291
Caracas 1060 A, Venezuela
Tel. 58-285-3111

Embassy of Venezuela, Commercial Section
1099 30th Street, N.W.
Washington, DC 20007
Tel. (202) 342-2214

Yugoslavia

American Embassy, Commercial Section
Kneza Milosa 50
Belgrade, Yugoslavia
Tel. 38-11-645-655

Bulletins

AID Procurement Information Bulletin
Advertises notices of intended procurement of AID-financed commodities. Available from AID'S US Office of Small and Disadvantaged Business Utilization/Minority Resource Center, Washington, DC 20523-1414. Tel. (703) 875-1551.

Country Marketing Plans (CMPs)
Prepared annually by the commercial sections of the American embassies for the U.S. Department of Commerce's U.S. and Foreign Commercial Service, covering over 60 countries.

The Exporter
Monthly magazine. Write to 4 West 37th Street, New York, NY 10018. Tel. (212) 563-2772.

Exporters Directory/U.S. Buying Guide – Journal of Commerce
110 Wall Street, New York, NY 10005. Tel. (212) 425-1616.

Exporter's Encyclopedia
Annual handbook covering more than 200 world markets. Available from Dunn's Marketing Services. Tel. (800) 526-0651.

Foreign Economic Trends
Published by the U.S. Department of Commerce. Each FET covers a single country. Available from Superintendent of Documents, U.S. Government Printing Office, Washington, DC 20402.
Tel. (202) 783-3238.

International Trade Reporter: Current Reports
Weekly newsletter. Available from Bureau of National Affairs, Inc.
Tel. (800) 372-1033.

World Factbook
Produced annually by the Central Intelligence Agency. Available from Superintendent of Documents, U.S. Government Printing Office, Washington, DC 20402. Tel. (202) 783-3238.

Glossary

Absolute ownership
The broadest form of estate in real property whereby the owner(s) possess all rights, title, and interest in the property.

Acceptance
A draft on which the debtor indicates by the word "accepted" his or her intention to pay or honor.

Acknowledgment
A standard form used by a vendor or supplier to advise that the purchase order has been received.

Act of God
Danger beyond human control; any accident produced by an irresistible physical cause, such as hurricane, flood, earthquake, or lightning. In no way connected with negligence.

Actual weight
Gross shipping or transport weight.

Addendum
A document used to modify bid documents prior to receipt of bids. An addendum is incorporated into the formal contract.

Ad valorem
The total value of goods or materials against which tariff rates are imposed.

Advertisement for bids
A published notice of an owner's intention to award a contract for construction work to a contractor who submits an acceptable proposal in accordance with the owner's Instructions to Bidders. In its usual form, the advertisement is published in a convenient form of news media (e.g., newspaper, magazine) in order to attract contractors who are willing to prepare and submit proposals for the completion of the project.

Advice of shipment
A notice sent to the purchaser from the seller advising that the shipment has gone forward; usually contains details of packing, routing, delivery date, etc.

After sight
A term used on bills of exchange meaning "after presentation to drawee for acceptance."

Agency
Implies a relationship between two parties in which one is empowered to perform certain functions or business transactions for the other.

Agency relationship
The relationship between a principal and agent.

Glossary

Agent
An agent is authorized by the principal to act in the principal's behalf and interest. An agent's actions generally bind the principal as though the principal had acted directly.

Aggregated shipments
An indefinite number of shipments from different supply sources to a single purchaser, consolidated and considered as a single consignment.

Agreement
A consensus by two or more parties. As it relates to construction, the term is synonymous with *contract*. An example of this is the agreement between owner and contractor.

Airbill
A shipping or manifest document used by airlines for air freight; contains shipping instructions to the airline.

Air freight
To transport or ship goods by air.

Airway bill
Document used for shipment and transport of air freight by air carriers; lists the materials shipped along with instructions, costs, and other specific details.

Allowance
A stated requirement of contract documents whereby a specified sum of money is incorporated, or allowed, into the contract sum in order to sustain the cost of a stipulated material, assembly, piece of equipment, or other element of a construction contract. This device is convenient in cases where the particular item cannot be fully described in the contract documents. The allowance can be stated as a lump sum or as a provisional sum.

All risk insurance
An insurance policy that covers specific risks of damage or loss from any number of potential events during and after the construction process.

Amortization
The distribution of the initial cost of an asset by periodic charges to ongoing operations, as in the case of depreciation.

Application for payment
A financial statement prepared by the contractor or subcontractor stating the amount of work completed and materials purchased and stored to date. The statement includes the sum of previous payments and current payments due in accordance with payment terms of the contract.

Approve
To accept and endorse as satisfactory; implies that the object approved has the endorsement of the approving agency or body. However, the approval may still require confirmation by another party.

Arbitration
The process by which parties agree to submit their disputes and claims to the determination and resolution of a third, impartial and unbiased party (referred to as the arbitrator), rather than pursuing their claims in a court of law.

Architect
A design professional who, by education, experience, and examination, is licensed by state government to practice the art of building design and technology.

Arrival date
The date purchased material and equipment are scheduled to arrive at the construction site.

Arrival notice
Notice that a freight carrier sends to the purchaser when a shipment has arrived.

Artisan's lien
The lien of a mechanic or other skilled worker in connection with something on which he or she has applied labor or materials, giving him or her the right to keep possession of it until final payment is made.

As built drawings
Record drawings made during construction. As built drawings record the locations, sizes, and nature of concealed items such as structural elements, accessories, fixtures, devices, valves, and mechanical equipment. These record drawings form a permanent record of the as built condition of the building or facility.

As is
Indicates that the materials and equipment offered for sale are without warranty or guarantee. The purchaser has no recourse on the vendor or supplier for quality of the materials and equipment.

Assignment
The transfer of rights or title to another party, frequently involving rights originating from a contract.

Attachment
A supplementary device designed to be fastened, fixed to, or mounted on a machine or implement.

Back order
The part of an order that cannot be delivered at the scheduled date, but will be delivered at a later date.

Banker's acceptance draft
A document or draft used in financing a foreign transaction, making possible the payment of cash to an exporter, covering all or part payment for a shipment made by the exporter.

Bargain
Agreement on the terms and conditions of a purchase. Purchase of articles at a price favorable to the buyer.

Barter
The process of exchanging one kind of article for another, as opposed to trading by use of money.

Bid
A complete and properly executed proposal to perform work or supply goods or services that have been described verbally or in the bidding documents and submitted in accordance with instructions to bidders. A bid is an offer.

Bid bond
A form of bid security purchased by a bidder; provided, subject to forfeit, to guarantee that the bidder will enter into a contract with the owner for construction of the facility within a specified time period.

Bidding documents
Documents that typically include the advertisement or invitation to bidders, instructions to bidders, bid form, form of contract, forms of bonds, conditions of contract, specifications, drawings and any other information necessary to completely describe the work for which bidders can prepare bids for the owner's consideration.

Bid opening
A formal meeting held at a specified place and time at which sealed bids are opened, tabulated, read aloud and made available for public inspection.

Bill
An invoice the freight carrier uses to show consignee, consignor, shipment description, weight, freight charges, and other relevant information. A statement of account, or money due for services rendered.

Bill of exchange
A formal written document used to settle and pay for an existing obligation.

Bill of lading
A transport company's contract and receipt for materials and equipment; agrees to transport from one location to another and to deliver to a designated individual or party.

Bill of materials
A list of all permanent materials required on the construction project. The bill of materials list consists of all items described on a drawing and specification.

Binder
A temporary but binding commitment by an insurance company to provide insurance coverage.

Blanket order
Provides for the vendor or supplier to furnish certain materials for a certain period of time and at a predetermined price; acts as a master purchase order, reducing the number of smaller purchase orders.

Boiler plate
A term used to describe the terms and conditions on the back of a purchase order or the specific clauses described in a contract.

Bonded warehouse
A warehouse under the surveillance of the U.S. Treasury Department for observance of revenue and excise laws.

Bond performance
A bond obtained in connection with a contract; ensures the performance and completion of all the scope, terms, conditions, and agreements contained within the contract.

Bonds
Formal documents, given by an insurance company, in the name of a principal to an obligee to guarantee a specific obligation. In the construction industry the main types of bonds are the bid bond, performance bond and payment bond.

Book value
Current investment value on a company's financial books calculated as original value less depreciation; the asset value of a building, facility, or item of equipment for accounting purposes. The value of an outstanding share of stock of a corporation at any one time, determined by the number of shares of that particular class that are outstanding.

Breach of contract
The failure to perform any of the obligations that are stated within the terms and conditions of the contract.

Bulk materials
Materials bought in lots; purchased from a generic description or standard catalog description and bought in medium to large quantity for issue as required. Examples are pipe fittings, conduit, cable, timber, and stone.

Bulletin
A document used to request pricing for a modification to the design after a contract is issued. If pricing is acceptable, a change order to the contract incorporates the requirements of the bulletin into the project.

Burden
In construction, the cost of operating a home branch or site office with staff other than operating site personnel. Also means federal, state and local taxes, fringe benefits, and other union contract obligations. In manufacturing operations, burden typically means operating overhead costs.

Cancellation of order
Annulment or a cessation of order.

Certificate of material compliance
A written statement signed and approved by an authorized person stating that the materials comply with the material specification.

Certificate of origin
A document, issued by the appropriate authority in an exporting country, that certifies the origin of the equipment, materials, or labor used in the manufacture of the equipment or materials being exported to another country.

Certified test report
A written document, approved by an appropriate body, that contains sufficient information to verify the actual properties of the materials and equipment and the actual results of the tests.

C. & F.
(Cost and Freight) Same as C.I.F., except marine insurance is not part of the selling price.

Change in scope
A change in requirements, objectives, work content, or schedule that results in a difference from the terms and conditions of the contract.

Change order
A written order, issued after execution of the construction contract, that authorizes a change in the construction work and contract time and/or value.

C.I.F.
(Cost, Insurance, Freight) When seller quotes C.I.F., the quote includes the cost of the materials or equipment, marine insurance, and all transportation charges to the stated destination point.

C.L.
Carload.

Claim
A request for additional payment.

C.O.D.
Cash on delivery.

Commercial terms
The terms and conditions of a purchase order or contract that relate to the business and commercial aspects of the purchase order or contract. The price, quantity, and delivery date are the main elements covered under the commercial terms.

Compensatory damages
Damages awarded to compensate the injured party by granting a monetary value equal to the loss or injury encountered.

Competitive bidding
The offer of proposals by individuals or organizations competing for a purchase order or contract to supply specific materials, equipment, or services.

Conditional sale
A sale made with the knowledge that title will not pass to the buyer until some stated condition has been achieved.

Conditions of the contract
A document describing the rights, responsibilities, and relationships of the parties to a contract.

Consequential damages
Payment for loss or damage that is not directly attributable to a wrongful action on the part of another party, but is the result of one or more of the consequences of the action.

Consideration
A term used to describe the value that shall be reimbursed to one party to a contract by another party in return for services or articles rendered.

Containerization
The use of road and marine transportation containers; containers are normally 20 or 40 feet in length. Shipment of large sealed freight containers via rail, air, truck, or water to optimize transit time, security, packaging, and turnaround time.

Contract
A written or oral agreement between two or more competent parties to perform a specific act or acts enforceable by law.

Contract administration
Administering contracts and purchase orders to protect the interests of a specific organization and to satisfy the conditions and requirements of the contract and/or purchase order.

Contract documents
A term applied to a collection of related documents (contract, specifications, drawings, and any additional data) that define the extent of an agreement between two or more parties.

Cost plus
A contract or pricing method in which the purchaser agrees to pay the supplier an amount determined by the actual costs incurred by the supplier to provide the materials, equipment, or services purchased, plus a fixed percentage of that cost or a fixed sum as profit.

Counteroffer
To decline an offer by submitting a new offer with different conditions or terms than the original offer.

Damage claim
A formal claim document filed for damages to material and equipment.

Damages
Compensation or payment for damage to materials and equipment, individuals, or property that is the fault or cause of another party.

Davis-Bacon Act
An act by the Congress of the U.S., passed into law in the 1930s. The Act provides that wages and fringe benefits paid to workers employed by contractors and subcontractors under contract with the federal government

be paid not less than the local prevailing wage rate for each trade.

Debit notice
An invoice used to offset a previous overpayment, showing the difference between the previous invoice and the correct value.

Deed
A legal document used to transfer ownership.

Deliverable
A product or report that must be delivered to satisfy a contractual obligation.

Delivering carrier
The carrier that transports and delivers the materials and equipment to the purchaser.

Delivery
The act of transferring possession; applied to shipping, occurs when lading is surrendered and title, materials, and equipment pass to the receiving party.

Demurrage
A charge made on freight cars, vehicles, or ships held by or for consignor or consignee for subsequent loading or unloading.

Detail specification
A description of the requirements for a specific item of material or equipment.

D.F.
Damage free.

Discount
An allowance or deduction given by the seller to the buyer that reduces the cost of the item purchased when certain conditions are met by the buyer (e.g., prompt payment within a stipulated period).

Distribution
The broad range of activities targeted at the efficient movement of finished materials and equipment from the end of the production line to the eventual end user or consumer.

Dock
(1)The loading or unloading ramp or platform at an industrial facility or factory.
(2) Pier, jetty or wharf for the receiving and embarking of ships.

Draft
A legal document instructing one individual to pay another.

Due date
The date when purchased materials and equipment will be available for installation at the project location.

Dunnage
Protective matter used around materials or equipment to prevent movement, damage, or breakage while in transit.

Durable goods
Consumer products that are used repeatedly over a period of years (e.g., household appliances, vehicles).

Duty
The charge assessed by a government on materials and equipment imported or exported. An obligation established by law or contract.

Earnest money
Money that one party gives to another at the time of entering into a contract to "seal the deal". Earnest money can be forfeited if a contract is not formalized.

Engineer
A design professional who by education, training, and experience is knowledgeable and skilled in the art and technology of engineering and design.

Engineers Joint Contract Documents Committee (EJCDC)
A committee formed in the 1970s that drafted standard forms of contract for use on engineering projects.

Errors and omissions excepted (e and o.e.)
Printed on invoices or other statement(s) to safeguard the originator's right to amend or modify the value if found to be incorrect.

Escalation
The value of adjustment permitted by an escalation clause. An allowance for an anticipated increase in the cost of equipment, materials, and labor as a result of continuing price inflation experienced over time.

Escalation clause
A contract clause that provides for a price adjustment based on specific changes.

Ex
(Ex Mill, Ex Factory, Ex Warehouse, Ex Dock) Prefix used to denote point of origin. When a seller quotes a price Ex, the seller proposes only to make the materials and equipment available at the Ex point of origin and includes no transportation costs in the quoted prices.

Excess freight
Freight in excess of that indicated on the original freight carrier billing.

Exchange bill of lading
A bill of lading compiled and exchanged for another bill of lading.

Excise tax
A tax on the manufacture, sale, or use of certain articles made, sold, or used within a country.

Export
Shipment of materials or equipment to a foreign country.

Export permit
A permit given by the government of an exporting country allowing a party within that country to export the materials and equipment to another county.

Fabrication
Manufacturing operations for materials or equipment as opposed to the final installation operation.

Facilitation
A system to decrease the time of international cargo transportation through the use of the latest customs methods, duty and tariff collection, and other related functions of international traffic activities.

Factor
An agent selling goods or materials on a commission basis for his or her principal.

Fair market value
The value of an article as determined by negotiation between a buyer and seller; considered acceptable as a basis of a purchase and sale of the particular article.

F.A.S.
(Free along side) When a seller quotes a price F.A.S., the price includes the cost of transportation and delivery alongside the oceangoing vessel and within reach of the vessel's loading equipment. The price does not include costs for any export permits or the payment of any export duties or tariffs.

Field inspection
A thorough examination of the equipment and materials shortly after delivery to determine if they meet the requirements of the specifications and to find any hidden defects or damage.

Field purchase order
A purchase order used in field construction situations where authority to make the type of purchase involved is usually restricted or has a predetermined not-to-exceed value.

Field required date
The date required for an item of material or equipment to be delivered to a construction site for the maintenance of the project schedule.

FIFO
An accounting procedure based on a first-in, first-out treatment of stock or inventory; the articles that are received earliest are used first. The opposite of this procedure is LIFO (last-in, first out).

Firm offer
A definite proposal or offer to buy or sell some article on stated terms and conditions. Such an offer binds the proposal to a stipulated time period.

F.O.B. (Free on Board)
Indicates that the seller pays the transportation costs up to the delivery location indicated. For example, a contractor's purchase order frequently specifies delivery as "F.O.B. construction site" or "F.O.B. storage yard." If the purchase order stipulates that the seller is to pay the delivery costs to a designated location, the title to the article does not pass until the article has been delivered to that location. Under F.O.B. agreement, title goes to the purchaser when the carrier delivers the article to the location stipulated.

F.O.B. destination, freight prepaid
A term used in reference to the title passing to the buyer and to the freight cost being paid by the seller.

F.O.B. factory, freight allowed
Free on board factory, freight prepaid from point of origin.

F.O.B. origin, freight collect
A term used in reference to the title passing to the buyer. The buyer pays the freight costs, owns the goods in transit, and files all claims for any loss or damage while in transit.

F.O.B. origin, freight prepaid and charged
A term used in reference to the title passing to the buyer. Freight costs are paid by the seller and then collected from the buyer by adding the value of freight costs to the invoice.

F.O.B. shipping point
The location at which title to the articles passes from the seller to the buyer. The seller is liable for transportation costs and the risks of loss or damage to the goods up to the point where title passes to the buyer. The buyer is liable for such costs and risks after the passing of the title.

Follow-up record
Information used in the expediting and delivery from vendors and suppliers.

F.O.R.
Free on rails.

Force majeure
Circumstances beyond an individual's control; pleadable as an excuse for the nonfulfillment of a contract or purchase order.

Forward purchasing
The purchase of quantities exceeding the immediate requirement, i.e., in anticipation of any significant price increase or market shortage.

F.O.T.
Free on truck.

Freight forwarder
Some freight forwarder organizations act as agents on behalf of shippers in organizing the transportation of articles without handling any of the articles. Others act as freight carriers in consolidating small and mid-sized shipments and delivering them to the purchaser.

General conditions
Guidelines that define many of the rights, responsibilities, obligations, and limitations of authority of the owner and contractor, and include a general procedure governing the performance of the work. When organized under the Construction Specification Institute (C.S.I.) MasterFormat, they are described under 16 divisions.

General contractor
A construction organization whose main business is the construction of various buildings or facilities.

General liability insurance
A broad form of liability insurance covering claims for bodily injury and property damage.

Generic
A term used to generally describe a group, type, or class of materials and equipment, rather than name a specific trade name or source of manufacture.

G.N.P. (Gross National Product)
The total of a country's national output of goods and services at current market prices.

Goods received note
A document detailing all equipment and materials after they have been audited and checked for quantity at the receiving point.

Gross ton
2,240 pounds of weight.

Gross weight
The total weight of a shipment, including containers, packaging and miscellaneous materials.

Guarantee
A promise, pledge, or formal assurance given as a pledge that another's obligation or debt will be fulfilled.

Guaranteed maximum cost contract
A contract for construction in which the contractor's or subcontractor's compensation is stated as a combination of actual cost incurred, plus a fee, with a guarantee by the contractor or subcontractor that the total compensation will be limited to a specific stated amount.

Heavy-lift charge
A fee imposed by a transportation organization for lifting materials and equipment of excessive weight.

Hedging
A method of selling for future delivery whereby the parties protect themselves against potential loss.

Hold
The below-deck cargo storage space aboard ship; the cargo compartment of an aircraft.

Hold order
A purchaser's order to hold a particular delivery at a designated location for a specific period of time.

Hold points
Inventory or warehouse areas set aside for storing semicomplete articles.

Hundredweight
In U.S. measurement and domestic transport, 100 lbs.; in United Kingdom measurement, a hundredweight, or "cwt" is 112 lbs. or one-twentieth of a long ton of 2,240 lbs.

Implied contract
A contract formed when parties express, through their conduct, their agreement to be bound to its conditions and terms. An agreement is inferred and understood without express statements.

Indemnification
An obligation contractually taken on or legally imposed on one party to protect another party against any loss or damage from specific liabilities.

Indemnity
A responsibility of one person to make good a loss or damage incurred by another. A payment for damage, loss, or expense incurred.

Indirect materials
A group of materials used in making a product that are not incorporated or part of the finished product.

Injunction
A formal order issued by a court of law ordering a person or a group to do or refrain from doing some activity.

Inland bill of lading
A bill of lading used in transporting materials and equipment overland to the exporter's international transporter. A through-bill of lading can be used in some circumstances; however, it is usually necessary to prepare both an inland bill of lading and an ocean bill of lading for foreign export of materials and equipment.

Inland carrier
A transportation organization that moves export or import materials and equipment between seaports and inland locations.

Inquiry
A request for information related to the schedule, location, availability, interest, cost, or quantity of construction-related items.

Inspection
The examination, audit, measuring, and testing of materials and equipment including, when necessary, raw materials, fabrication elements, components, intermediate assemblies, subassemblies, and end products to determine if the materials, equipment, and services meet the contract specification requirements. This activity can be completed either by visual inspection or with the use of special equipment.

Inspection plan
A procedure that defines the material or equipment requiring inspection and describes the method and order of performing the tests or inspections.

Inspection report
A report to inform the purchaser of the quality and

workmanship of the items of material and equipment delivered.

Inspector
A qualified inspector employed by an organization to examine and review the fabrication, manufacture, and installation of construction-related equipment and materials.

Instructions to bidders
A document that is part of the bidding requirements; usually prepared by a design professional, architect or engineer. Instructions to bidders describe specific instructions to the potential contractors on procedures, requirements of the owner and other necessary information for the preparation and submission of bids for consideration and review by the owner.

Insurance
A contract, typically called an insurance policy, in which the insurer, in return for the payment of a premium, agrees to pay the insured for any losses or damages incurred by the insured during the execution of a specified undertaking (e.g., construction of a building or facility).

Insurance certificate
A document supplied by a contractor, subcontractor, or vendor stating the coverage of insurance related to a particular contract.

Inventory
Items of material and equipment that are in the storeroom or warehouse, or work-in-progress consisting of raw materials, fabrication elements, components, parts of intermediate materials and equipment, and finished materials and equipment ready for distribution and sale. Physical inventory is ascertained and established by actual count. It includes materials and equipment that are physically available for allocation and distribution, and stored and controlled in a warehouse or outside laydown area.

Invitation to bid
Written notice of an owner's desire to receive competitive bids for a particular construction project where a select number of contractors are invited to submit bids for the construction of the project.

Invoice
Seller's itemized bill of quantities and prices of materials, equipment, and/or services that have been delivered to the purchaser.

I.T.C.
Investment tax credit.

JIT (Just-in-time)
A system for planning and scheduling fabrication through the manufacturing sequence, optimizing the purchase and storage of materials and equipment, and eliminating any inventory or stock on hand.

Job lot
A small number of specific materials or items of equipment that are produced at one specific time.

Job-lot ordering
The buying of necessary materials, equipment, and components to fabricate in accordance with a customer's specifications.

Joint agent
An official designated to act for two or more principals.

Joint venture
An organization in which two or more parties join together to form a business operation with the legal characteristics of a partnership to achieve a specific goal.

Journeyman
A qualified workman who has completed an apprenticeship.

Judgment
A decision rendered as a result of a course of action in a court of law.

K.D.
Knocked down or disassembled to reduce bulk.

Key Activity
An activity that is considered of major importance, sometimes referred to as a milestone event.

Kitting
The process of sending components of a total assembly to another location in a kit form for assembly.

Labor and material payment bond
A contract between a contractor and a surety in which the surety, for a premium payment by the contractor, agrees to reimburse subcontractors, vendors, and material suppliers any amounts due for their materials and services, should the contractor default in the payment to them.

Lading
The act of loading, or the contents of a specific shipment.

Laydown area
A space of ground, usually without a roof, that is used for the delivery and storage of materials and equipment.

L & D
Loss and damage.

Lead time
The period of time required to perform a specific activity of work.

Lease
A contract whereby an individual or organization lets another use property or equipment for a definite term and for an agreed rental cost. The lessor retains title to such property or equipment.

Legal tender
Money issued by the government to satisfy a debt or obligation.

Letter of credit
A letter addressed by a bank to a correspondent bank certifying that an individual or organization named therein is entitled to draw upon an account.

Liability
A state of being under obligation; exposure to potential claim by which an individual or organization may be subject to pay compensation for loss, damage, or other acts to another individual or organization.

Lien
A legal claim on the assets or property of another.

LIFO (Last in, first out)
An accounting practice of determining the cost of stock inventory used in a manufactured product or process.

Lighter
An open or covered barge used mostly in harbors or ports for transferring materials and equipment between vessels and docks.

Liquidated damages
A sum of money agreed to by the contracting parties as to damages to be given in case of a failure to meet the obligations of the contract.

Litigation
To engage in a lawsuit; the process by which parties submit their disputes to the jurisdiction of federal or state law courts for resolution.

Load factor
A ratio that applies to the utilization of a physical plant or piece of equipment; the ratio or percentage of average load utilized to maximum load available for use.

Logistics
The science of transportation and supply; the art of obtaining and distributing finished products in the marketplace.

Long ton
2,240 pounds; same as gross ton.

Lot size
The amount of specific materials and equipment items ordered from a vendor.

L.T.L.
Less than truckload.

Lump sum
An amount or value used in a proposal, bid or contract representing the total cost that an organization is prepared to contract to perform an item of work.

Manifest
A list of the cargo loaded in ships, trucks, containers, etc.; on oceangoing transportation, referred to as a ship's manifest.

Marketable title
A title about which no uncertainty exists concerning its legal soundness or validity.

Marketing research
The systematic gathering, recording, and analyzing of data and intelligence about tasks and activities related to the marketing of materials and services.

Markup
A percentage that can be added to the total of all direct costs to determine a final price or contract sum. Allows the contractor or subcontractor to recover the costs associated with overhead.

MasterFormat
A 16-division numerical system of organization for construction-related data, developed by the Construction Specifications Institute (CSI) of the U.S. and the Construction Specifications Canada (CSC).

Material
The raw elements, parts, or semiprocessed components from which a finished product is created.

Material cost
The cost of all items that are essential to the construction process or operation of a facility, including the direct and indirect related costs.

Materialman
A term used to describe an individual from whom the contractor may obtain the materials of construction. The materialman may be a manufacturer's sales representative or agent or distributor; a salesman of materials and equipment could also be considered a materialman.

Materials management
A concept whereby all materials and equipment procurement functions are combined under one management function, including contracting, purchasing, quality assurance, quality control, inspection and expediting, trafficking, and receiving.

Material status
A report detailing the current availability of materials.

Materials test report
A report usually referred to as a Certified Materials Test Report or Mill Test Report. The actual test results are usually described, detailing chemical analyses, material composition tests, and procedures used in testing.

M.C.
Minimum charge.

Mechanic's lien
A type of lien filed by an individual or organization who has performed work for which payment is either in dispute or remains unpaid.

Minimum carload weight
A minimum weight for which a carload of materials and equipment can be charged.

Mixed truckload rate
A rate applied to a truckload shipment made up of two or more different materials.

Modifications
A term used to signify changes that may be made to contract documents and a construction contract. Modifications made before the award of a contract are called addenda; modifications made after the contract is in place are called change orders.

Multiple-consignee
A container car, truck, or ship loaded with materials and equipment for two or more consignees.

Multiple source buying
The procedure of finding new sources of materials and equipment.

Need date
The date when a particular article is required at the construction site.

Negligence
Under the law, failure to exercise the care and consideration a prudent person would exercise; lack of care and attention.

Negotiation
The process by which a buyer and seller reach an agreement on the terms and conditions regarding the purchase of materials, equipment, or goods.

Nest
A set or series, often from large to small, such that each article fits within another. This situation reduces the bulk and facilitates handling for storage or shipment.

Net price
The price reached after all allowable discounts, rebates, etc., are deducted from the original selling price.

Net ton
2,000 pounds.

Net weight
The weight of the materials and equipment without the shipping container and dunnage.

Notary
A public officer empowered to administer oaths, take depositions, and certify deeds and contracts.

O.B.L.
Ocean bill of lading.

Obligation
A duty that is the result of a promise or contract. An agreement which an individual or organization is responsible to fulfill.

Obsolete
Outmoded, worn out, discarded, or no longer in use.

Offer
A proposal or bid made by an individual or organization to another individual or organization to perform a service or action; the acceptance of such an offer results in a contract. The individual or organization who makes the offer is called an *offeror*, and the individual or organization who receives the offer is called an *offeree*. A bid or proposal is an example of an offer.

Open Competitive Bidding Selection
A process of contractor selection wherein an advertisement to bidders is published in newspapers and trade magazines notifying contractors of the owner's intention to receive and consider sealed competitive bids for a construction contract. Typically the lowest conforming and responsible bid will be successful.

Open-end order
Purchases made against the buyer's purchase order or contract. The purchase order or contract contains price, conditions, and terms. The purchase order or contract may not specify the final quantity to be purchased.

Order lead time
Period of time required to obtain an item from a vendor or supplier once the purchase order requirements are known.

O.S.& D. (Over, Short, and Damage Report)
A report or log showing discrepancies in materials received, together with a damage evaluation.

Overhead
A cost inherent in the operating of a business. A cost that cannot be charged to a specific part of the work, material, or equipment. These costs must be allocated as a percentage to the work, material, or equipment.

Owner
The individual, group, or organization that has title to a building or facility.

Packing list
A document or log prepared by the shipper to indicate in detail the particular package contents.

Partial payment
A stage payment made upon delivery of one or more completed units.

Patent
A government grant to an inventor by which he or she is the only person allowed to make or sell the new invention for a certain number of years.

Payment bond
A form of security purchased by a contractor from a surety; guarantees that the contractor will pay all costs of labor, materials, and other services associated with a construction project.

Penalty clause
A clause in a contract that stipulates the sum of money to be forfeited in the case of nonperformance of the terms and conditions of the contract.

Performance bond
A form of security purchased by a contractor from a surety; guarantees that the contractor will satisfactorily perform all

of the work associated with a construction project.

Physical distribution
The activities associated with the transportation and movement of materials and equipment from the manufacturer to the end user.

Piggyback
The carrying of anything that usually moves alone by a large vehicle. The transportation of highway trailers or containers on specially equipped railroad flat cars.

Pledge
To bind by a promise.

Point of origin
(1)The location at which a shipment is received by a transportation company.
(2)The actual location of origin of an article.

Port of entry
A port designated by a government as the entry point for material, equipment, and services from an overseas country.

Preassembly
A fabrication process by which various materials, components, and equipment are combined together at a location other than the construction site for subsequent installation at the construction site.

Prefabrication
A manufacturing or fabrication technique, generally taking place at a location other than the construction site, in which various materials and equipment are combined to form a larger component element for final installation at the construction location.

Procurement
The activity related to the acquisition of articles, land, property, or services by the means of purchasing.

Procurement lead time
The time required by a buyer to select and negotiate with a vendor or supplier and place a purchase order.

Promissory note
A written pledge or promise by one individual or organization to pay another unconditionally a certain sum of money (principal and interest) at a specified time.

Purchase
Obtaining an article for money; something acquired for a specific amount of money or its equivalent.

Purchase order
A written contract made between a buyer and seller that describes the articles being purchased, the price of the articles, and the method of delivery.

Purchase requisition
A written request issued to a purchasing department from an individual or group that requires a specific article.

Purchasing
The art of buying materials, equipment, and services that conform to the correct quality, in the correct quantity, at the market price, and are delivered in accordance with the promised delivery date.

Purchasing cycle
The activities in the acquisition of materials, equipment, and services.

Purchasing manual
An operating guide that explains the policies and procedures for purchasing personnel to follow in the performance of their work activities.

Quality
The essential attributes that permit materials and equipment to function in the desired manner.

Quality assurance
A formal procedure that ensures that the material and equipment will perform satisfactorily when installed.

Quality control
The procedure and activities that ensure adequate quality is maintained in the materials and equipment utilized in the construction process.

Quality surveillance
The observation and evaluation of the manufacturing process and subsequent installation of materials and equipment.

Quantification
A list of actual quantities of materials and equipment from a set of drawings and specifications.

Quantity
The amount of equipment or material units required.

Quantity discount
The reduction in unit price cost established by a predetermined minimum number.

Quasi-contract
An obligation under which an individual or organization that received a benefit must pay the individual or organization who gave the benefit, despite the absence of a contract.

Quotation
A summary of price, terms of sale, and general description of materials, equipment, or services offered for sale by a contractor or vendor to a potential buyer. When issued in response to a purchase inquiry, it is considered an offer to sell.

Quotation expiration date
The date after which a quotation is no longer valid.

Quotation request
A purchaser's invitation to potential vendors or suppliers to bid on a list of articles.

Rate of exchange
The rate at which the currency of one country is exchanged for the currency of another country.

Reactive expediting
A form of expediting conducted primarily by telephone contact; reacts to situations and problems as they occur.

Rebate
The amount refunded to a purchaser for the purchase of an agreed quantity or value within a stipulated period of time.

Receipt inspection
An audit and examination of materials and equipment prior to acceptance; to review the completeness of the delivery and to note any obvious damage.

Receipt of Bids
The action of an owner in receiving sealed bids that have been invited or advertised in accordance with the owner's intention to award a construction contract.

Receiving
The receipt and delivery of articles at a designated location.

Receiving and storage
The action of possessing materials and equipment from vendors and suppliers, including maintaining and controlling all elements in storage, and the eventual distribution of the materials and equipment.

Receiving report
A form or log used by an organization's receiving department for recording the materials and equipment received and the differences, if any, from the quantities indicated on the purchase order.

Recourse
The procedure a buyer or seller may use to have a seller or buyer satisfactorily comply with the terms of the contract.

Repair
To mend or put in good condition. The process of restoring a nonconforming material or equipment item so that it can perform its intended function.

Request for bid
A request by a purchaser for the submission of an offer or proposal from the supplier to be considered, accepted, or rejected by the purchaser.

Requirements
The total scope of work to be performed together with specifications, drawings, and cost of the work.

Requisition
An internal form that an individual or department sends to the purchasing department requesting materials, equipment, or other services.

Retention
The percentage or value withheld from a contractor that is paid when the contract has been satisfactorily completed.

Review
The examination of any form of data documentation drawings, for the purpose of establishing acceptability and conformance to the requirements of the contract.

Rework
The procedure by which a nonconforming element is made to conform to the established need by repairs or modifications.

R.F.Q. (Request for quotation)
The process of soliciting bids or proposals for a stated scope of work.

Ro/Ro (Roll-on/Roll-off)
A specially constructed ship that permits road vehicles to drive on and off the ship.

Routing
The determination of each road to be used in the transportation of materials and equipment to a construction job site.

Royalty
A fee that is payable to the owner of a patent.

Sales price
The value received for items sold. Gross sales price is the total value paid. Net sales is the gross sales value less discounts, rebates, and freight costs.

Sales representative
An individual acting for a supplier or vendor who is familiar with the products that the buyer may be considering for purchase.

Sales revenue
Funds received as the result of a sales transaction.

Sales tax
A tax levied on the sale of materials, equipment, or services; calculated as a percentage of the purchase price. Different states use different percentages.

Salvage
The material or equipment that is saved after damage or demolition has been completed.

Salvage value
The value recovered or realized when an article or facility is demolished, scrapped, or sold.

Samples
Examples of completed materials, products, equipment, or workmanship that establish standards by which the installed work will be evaluated.

Scope
Defines the work to be performed and completed; usually documented and described in the contract.

Scope change
A deviation from the original project scope of work agreed to in the contract. A scope change can be an activity either added to or deleted from the original scope of work.

Seller's lien
The seller's right to withhold or to lay claim on materials or

equipment sold, giving up these rights upon receipt of payment.

Shop drawings
The various drawings created by contractors, subcontractors, vendors, or manufacturers, that illustrate construction, materials, dimensions, and installation data for the incorporation of a specific element into the construction project.

Shop fabrication
The manufacturing assembly of components or elements in a vendor's or manufacturer's shop.

Shop inspection
A detailed inspection and audit at the vendor's or fabricator's shop of the conformance of materials and equipment to the project specifications.

Short ton
2,000 pounds.

Small tools
In construction, a saw, shovel, hammer, trowel, etc., that is operated by hand.

Sourcing
The process of researching and determining qualified sources of materials and equipment.

Specifications
A precise, detailed description and presentation of some article.

Standard
The requirements of a specific standardization method approved by a recognized and appropriate authority.

Start-up
The systematic audit and check-out of all plant equipment and systems in accordance with prescribed procedures and tests, commencing prior to completion of construction and extending through mechanical completion.

Status report
A type of log or schedule report, chart, or graph, that monitors planned and actual physical progress.

Stop order
A document used to direct a work stoppage. Contractor, subcontractor, or vendor is required to acknowledge receipt of a stop order.

Storage
The function of placing materials and equipment in a designated area for safekeeping and distribution.

Subcontract
An agreement between a general contractor and a contractor who specializes in a specific trade (e.g., roofing, electrical) for the performance of a portion or element of work for which the general contractor is responsible to the owner.

Submittal
A sample, manufacturer's cut sheet or data, shop or fabrication drawing or other item that is submitted to the owner, architect, or engineer by a contractor for approval or other action.

Subsystem
A group or set of assemblies, components or elements, that when combined perform a single function or purpose.

Supplier
Vendor, seller, manufacturer, contractor, or subcontractor.

Supplier evaluation
The procedure of evaluating a supplier's ability to perform the required quantity, quality, and schedule requirements.

Surety
An individual or company that issues a bond to guarantee that another person or company will perform in accordance with the terms and conditions of an agreement or contract.

Surplus
The usable materials, equipment, components or parts that are in excess of the construction requirements.

Surveillance
The monitoring and witnessing of construction work.

Surveillance inspection
The observation and inspection of the manufacture of materials and equipment used in construction.

Tagged item
A separate identifiable item, generally tracked and controlled separately from the bulk materials or commodities; an example of this is the instrumentation items.

Takeoff (or quantity takeoff)
The process of compiling the required material quantities from the contract drawings.

Tariff, freight
A listing of duties or taxes on imports or exports.

Tax exemption certificate
A document given by the purchaser to the seller with the purchase order to indicate that the transaction is not subject to state sales tax.

Technical bid evaluation
The ranking of vendor or supplier bids based on the quality, cost, compliance with specifications, and delivery requirements.

Terms of payment
The method of payment for materials, equipment, and services stipulated in a contract.

Testing
Confirming an article's ability to meet pre-established requirements by subjecting the article to a set of physical, chemical, or operating evaluations.

Title
The legal right to the possession of property.

Tracing
Locating the current position of a shipment after it has entered the delivery and transportation phase.

Trademark
A mark, picture, name, or letters owned and used by a manufacturer or merchant to distinguish his or her goods from the goods of another.

Traffic
The action of transporting materials and equipment by a freight carrier.

Trailer
A vehicle design without motor or power, to be drawn by another vehicle.

Transit
Action of being conveyed from one location to another.

Transit charges
Costs of services rendered while a shipment is being transported.

Transit rate
A rate applying to traffic stopped en route for milling, painting, packing, treating, storage, etc.

Transload
Shipment stopped while being transported in order to be partially unloaded.

Transmittal
A form or letter indicating the action to be taken on an article being transmitted from one party to another.

Trial
The term commonly applied to an action in a court of law; the examining and deciding of a case in court.

Turn key
A form of contract that provides all of the necessary services to complete a building, facility, or other construction project.

Unit cost contract
A construction contract in which reimbursement is based on a pre-established cost per unit of measure for the quantities produced or installed.

Unit load
Several articles that are loaded on one pallet, or placed in a crate, enabling transportation of the items at one time, as one unit.

Unit of measure
Used to specify the number of units or items to be purchased.

Unit train
Freight trains that move large quantities of bulk materials between two or more locations.

Usage
The number of units or articles of an inventory item consumed over a period of time.

Use tax
A tax imposed on the user of material and equipment.

Valuation
The appraisal of the value of exported or imported materials and equipment.

Value
The real worth of an article; marketable price. Intrinsic worth of an item. The value of an article is determined by the lowest cost at which a satisfactory supply of materials and equipment or services can be obtained.

Value analysis
The application of techniques that establish a value for a necessary action at the lowest evaluated cost.

Value engineering
A discipline that reviews the real value of various life cycle costs, materials, equipment and construction techniques. Value engineering reviews the initial cost of design construction, coupled with the costs associated with maintenance, energy use, and life cycle.

Variable costs
The costs associated with raw materials and all manufacturing operating costs, which vary with manufacturing output (e.g., water, electricity, gas and catalysts).

Variance
The permission granted to an owner of land or property to depart from the requirements of a specific zoning ordinance. A variance allows the owner of land or property to use his or her land in a different manner than is specified in the original ordinance.

V.A.T.
Value added tax.

Vendee
The buyer of materials, equipment, or services; a purchaser.

Vendor
An individual or organization that sells something to a purchaser.

Vendor performance evaluation
A ranking and evaluation of vendors' and suppliers' performance based on quality of work, compliance with specifications, delivery, and cost.

Vendor's lien
A seller's right to retain possession of materials or equipment until he or she has received payment.

Verification
Witnessing of certain steps in the fabrication and manufacturing process, such as metallurgical analysis, hydrostatic and performance or operational tests. Review and audit of non-destructive testing, x-rays, and bench tests.

Vessel storage plan
A plan of the arrangement of freight containers aboard a vessel or ship, showing the physical location of each container.

Vessel-ton
One hundred cubic feet of volume.

Vicarious liability
An individual or organization who is liable for the results of another individual's or organization's failure to act.

Visual inspection
Manual inspection of materials and equipment.

V.N.X.
Value not exceeding.

Volume discount
A reduction in unit cost predicated on the size of a particular purchase.

Waive
To give up a right or claim; to refrain from claiming or pressing.

Warehouse
A facility for the receiving, storage and eventual distribution of materials and equipment.

Warehouse receipt
A document given by the warehouseman as a receipt for materials and equipment placed in the warehouse.

Warranty
A promise or pledge that something is what it is claimed to be. The seller makes a specific assurance concerning the nature, quality, and character of the goods.

Waste
The refuse from the fabrication and manufacturing process, that cannot be reclaimed or reused.

Waybill
A document compiled at the point of origin of a shipment indicating the point of origin, final destination, route, purchaser, seller, description of shipment, and the cost of the transportation.

Weight, gross
The actual combined weight of the item, container, and any dunnage materials.

Weight, net
The actual weight of the item; does not include container, or any dunnage materials.

Weight, tare
The difference between the gross weight and the net weight of an item being shipped; weight of empty container, including dunnage packing material used in transport.

Wharfage (WHF)
The cost required by a pier or dock operator for freight movements carried out at the pier or dock.

WHF
Wharfage.

Wholesaler
An individual or organization that acquires products, materials, and equipment for resale to retailers or other users.

Witness
The task of observing and monitoring; to watch over; to examine; specific tests or work activities that may include a sign-off procedure.

Workers' compensation insurance
The insurance required of employers that provides compensation for injury and loss of wages for a work-related accident.

Work in process
Materials or equipment in various states of fabrication or completion.

Work Order
A written form that describes an activity to be completed.

Index